FASTtrack

Chemistry
of Drugs

FASTtrack

Chemistry of Drugs

David Barlow

Reader, Pharmacy Department,
King's College London, UK

David Mountford

Lecturer, Pharmacy Department,
King's College London, UK

Pharmaceutical Press
London • Chicago

Published by the Pharmaceutical Press
1 Lambeth High Street, London SE1 7JN, UK

© Pharmaceutical Press 2014

 is a trade mark of Pharmaceutical Press

Pharmaceutical Press is the publishing division of the Royal Pharmaceutical Society
of Great Britain

First published 2014

Typeset by Laserwords Private Limited, Chennai, India
Printed in Great Britain by TJ International, Padstow, Cornwall

ISBN 978 0 85711 083 1

A catalogue record for this book is available from the British Library.

Contents

Introduction to the *FASTtrack* series

FASTtrack is a series of revision guides created for undergraduate pharmacy students. The books are intended to be used in conjunction with textbooks and reference books as an aid to revision to help guide students through their exams. They provide essential information required in each particular subject area. The books will also be useful for pre-registration trainees preparing for the General Pharmaceutical Council's (GPhC) registration examination, and to practising pharmacists as a quick reference text.

The content of each title focuses on what pharmacy students really need to know in order to pass exams. Features include*:
- concise bulleted information
- key points
- tips for the student
- MCQs and worked examples
- case studies
- simple diagrams.

The titles in the *FASTtrack* series reflect the full spectrum of modules for the undergraduate pharmacy degree.

Titles include:
Applied Pharmaceutical Practice
Complementary and Alternative Medicine
Law and Ethics in Pharmacy Practice
Managing Symptoms in the Pharmacy
Pharmaceutical Compounding and Dispensing
Pharmaceutics: Dosage form and design
Pharmaceutics: Drug delivery and targeting
Pharmacology
Physical Pharmacy (based on Florence and Attwood's *Physicochemical Principles of Pharmacy*)
Therapeutics

Additional questions are available at www.ontrackpharmacy.com by selecting the FASTtrack option.

If you have any feedback regarding this series, please contact us at feedback@fasttrackpharmacy.com.

*Note: not all features are in every title in the series.

Preface

This text will undoubtedly be useful for stabilising the rogue leg of that annoyingly wobbly 1960s coffee table in the "lounge" of your shared student house. It could also be profitably employed to wedge the toilet door shut now that the locking bolt has fallen off, or even find use as an aid to sleep when you find yourself suffering insomnia. Its more prosaic use, however, but yet the use for which it was lovingly crafted by its authors, is as a revision aid for Pharmacy students who are preparing for examinations involving organic chemistry, elementary medicinal chemistry, and biochemistry.

The book should be seen neither as a substitute for lecture notes, nor as a cheap alternative to the more comprehensive textbooks. Rather, it should be viewed as a valuable adjunct to your own lecture notes, and as a resource for you to gauge your progress in learning as you swot up on the chemistry of drugs.

A detailed knowledge and understanding of the chemistry of drugs is fundamental to the discipline of Pharmacy. Such knowledge and understanding allows the practising Pharmacist to appreciate the methods by which drugs are synthesised, the ways in which they are analysed and tested prior to licensing and marketing, and the ways in which they are quality assured in manufacture. It is the chemistry of the molecules that is responsible for their pharmacological activity; it is their chemistry that determines the ways in which they are formulated as medicines; and it is their chemistry that determines their stability on the shelf in the home, and their stability and fate within the human body.

When you do delve inside this book, you might first like to revise one of the topics (using your own notes) and then test yourself on that topic by attempting the self-assessment multiple choice questions given at the end of the relevant chapter. That way, you'll get some idea as to the quality of your notes and/or revision of that topic. You can then read and digest the material presented in that chapter, hopefully improving your knowledge and understanding of the subject as you do so.

David Barlow
David Mountford
March 2014

About the authors

DAVID BARLOW is a reader in computational & molecular biophysics and head of the pharmaceutical chemistry teaching section in the Pharmacy Department at King's College London. He is a biochemistry graduate with an MSc in bio-molecular structural organisation and a PhD in crystallography.

His research over recent years has been primarily concerned with structural and computational studies of drugs and drug delivery systems. These studies have included: informatics and virtual screening of phytochemical libraries; development of expert systems approaches for drug discovery and modelling of drug delivery; development of novel software for modelling membrane structure; detailed characterisation of the molecular architectures and membrane interactions of drug and gene delivery vehicles; and the development and exploitation of heterologous expression systems for production of integral membrane proteins and novel peptides for use in pharmaceutical formulation.

DAVID MOUNTFORD is a lecturer in medicinal chemistry at King's College London, having previously worked in the pharmaceutical industry as a senior medicinal chemist working on a range of indications with an unmet medical need.

His current research is focused on the development of new methods for the preparation of classes of compounds that are either poorly described or not described at all in the literature. A second strand of research lies in the application of drug discovery methods to biological targets: both those that are well described and potentially new drug targets arising from original research being carried out at King's College London. In addition to using well established methods, he is working on the development and evolution of existing drug discovery methods and their application to those targets traditionally considered to be less druggable or undruggable.

chapter 1
Chemical structure and bonding

Overview

After learning the material presented in this chapter you should:
- understand how an atom is structured, in terms of both the nucleus and the electronic orbitals
- understand the assignment of an electronic configuration to atoms and ions
- be able to discuss how atomic orbitals combine to form molecular orbitals
- appreciate how sp^3, sp^2 and sp hybrid orbitals are formed
- be able to discuss how hybridised atoms form single, double and triple bonds
- understand the factors that determine molecular shape and bond angles, and appreciate why some molecules deviate from theory
- review the drawing of resonance hybrids using the curly arrow notation.

- An atom is composed of a nucleus and a surrounding cloud of electrons.
- The nucleus contains both protons and neutrons and has an overall positive charge.
- Fundamental properties

Particle	Proton	Neutron	Electron
Charge	+1	0	−1
Mass (amu)	1.00728	1.00866	0.00054

- The atomic number of an element is given by the number of protons in the nucleus or the number of electrons in the unionised element.
- The mass number of an element is given by the sum of the protons and neutrons (Figure 1.1).
- Atoms which contain the same number of protons in the nucleus but a different number of neutrons are termed isotopes. Isotopes have different physical and chemical properties.

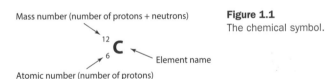

Mass number (number of protons + neutrons)

$^{12}_{6}\text{C}$ Element name

Atomic number (number of protons)

Figure 1.1
The chemical symbol.

Cations and anions

■ An atom can be ionised by either adding or removing an electron. The addition of an electron results in the formation of a negatively charged species, an anion. The removal of an electron results in the formation of a positively charged species, a cation.

Electron orbitals

■ The electrons of an atom occupy discrete atomic orbitals. These orbitals describe the probability of locating an electron in the region outside the nucleus.
■ In the same way that light demonstrates both wave and particle-like behaviour, electrons also demonstrate this duality, and hence an electron can be described by its mathematical wave function.
■ Individual atomic orbitals can accommodate a maximum of two electrons. These must be of opposite spin and are termed spin-paired.
■ Atomic orbitals can be described as s, p or d orbitals (Figure 1.2).
■ The three p orbitals are degenerate; they are each of the same energy.
■ The five d orbitals are degenerate; they are each of the same energy.

Figure 1.2
Shapes of s, p and d atomic orbitals.

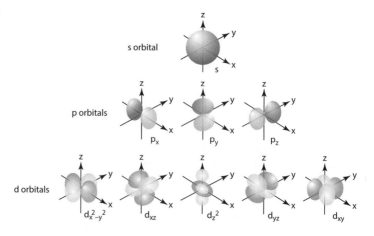

Electronic configuration

■ Atomic orbitals (Figure 1.3) are filled to give the lowest-energy electronic configuration for an atom, the ground state.

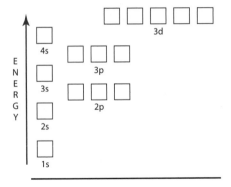

Figure 1.3 Energy level diagram.

- The ground state electronic configuration of oxygen can be found by accommodating eight electrons (Figure 1.4).

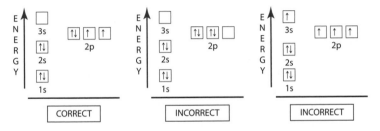

Figure 1.4 Correct and incorrect filling of atomic orbitals for oxygen.

- When a shell of electrons becomes full it is termed core. Core electrons typically do not participate in chemical reactions.
- Electrons in incomplete shells are termed valence. Valence electrons are available to participate in chemical reactions and bonding.

Molecular orbitals

- In the same way that electrons in an atom occupy atomic orbitals, the electrons in a molecule occupy molecular orbitals.
- Molecular orbitals are formed from the linear combination of the individual atomic orbital wave functions. This combination can either be as the sum of the two wave functions or as the difference of the two wave functions. That is,

$$\Psi = c_a \psi_a + c_b \psi_b$$
$$\Psi^* = c_a \psi_a - c_b \psi_b$$

Tips

- The bonding molecular orbital is lower in energy than the initial atomic orbitals.
- The antibonding molecular orbitals are higher in energy than the initial atomic orbitals.

where $c_a\psi_a$ and $c_b\psi_b$ describe the individual atomic orbitals, Ψ describes the bonding molecular orbital and Ψ^* the antibonding molecular orbital.

- The linear combination of two atomic orbitals will result in the formation of two molecular orbitals, one of which will be bonding, the other antibonding (Figure 1.5).

Figure 1.5
Forming molecular orbitals from individual atomic orbitals.

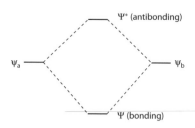

- If the two atoms are identical, each will contribute equally to the molecular orbitals. The bonding molecular orbital will be lower in energy than the starting atomic orbitals by the same amount as the antibonding molecular orbital is higher in energy than the starting atomic orbitals.
- Each molecular orbital can accommodate a maximum of two electrons, each with opposite spin.
- Molecular orbitals are filled with electrons in the same way as atomic orbitals, starting at the lowest energy available molecular orbital and spin pairing before filling the next highest molecular orbital.
- In a hydrogen molecule (H_2) each hydrogen atom is contributing a single 1s electron; these are spin paired in the lowest energy molecular orbital, the sigma (σ)-bonding orbital. The σ^*-antibonding orbital remains empty (Figure 1.6).

Tips

If you bring the two hydrogen 1s orbitals together so that they overlap you can imagine a region of increased electron density between the two atoms. This is like adding the two hydrogen wave functions together to create the bonding molecular orbital. If you now consider the region of overlap and subtract one of the overlaps from the other you would end up with nothing – a region without electron density between the two atoms. This is what the antibonding molecular orbital looks like.

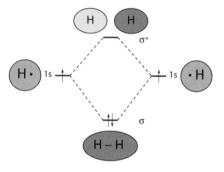

Figure 1.6 Molecular orbitals for molecular hydrogen.

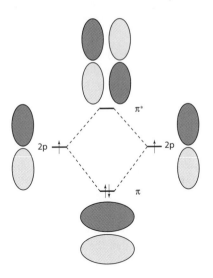

Figure 1.7
Molecular orbitals from
p orbitals.

- In the same way that s orbitals combine to give σ and σ* molecular orbitals, p orbitals combine to give pi (π) and π* molecular orbitals (Figure 1.7).
- Although s orbitals and p orbitals describe individual atomic orbitals, they are not suitable for describing molecular orbitals.

Promotion

- Promotion allows an element to increase the number of bonds it can make through the movement of a paired non-core electron to an empty orbital.
- In the case of carbon, paired 2s electrons would only allow the formation of two covalent bonds. Promotion of one of these electrons to the empty 2p orbital allows all four electrons to participate in bonding (Figure 1.8).

KeyPoints

- All group 2, 3 and 4 elements promote an electron in this way to increase their bonding capability.
- Phosphorus and sulfur are able to promote into nearby vacant d orbitals to form five and six bonds respectively.

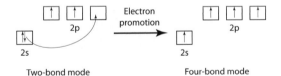

Figure 1.8 Electron promotion in carbon.

Hybridisation

- The individual atomic orbitals on an atom must be combined in a process termed hybridisation so that the resulting molecular orbitals are of the correct geometry and type for bonding to occur.

sp³ hybridisation

- In sp³ hybridisation the 2s and all three 2p orbitals are combined to create four sp³ hybrid orbitals.
- Each hybrid orbital has a 1/4 contribution from each of the four starting atomic orbitals.
- The four sp³ hybrid orbitals are of an equivalent energy, which is lower than the starting 2p orbitals but higher than the starting 2s orbital.
- During hybridisation orbitals are not created or destroyed. The four atomic orbitals mix to give four hybrid orbitals.
- Electrons in hybrid orbitals experience electron–electron repulsion in the same way as electrons in atomic orbitals. Each sp³ hybrid orbital will be occupied singly before spin pairing if necessary (Figure 1.9).
- The four sp³ hybrid orbitals adopt a tetrahedral geometry to minimise electron–electron repulsion between the hybrid orbitals (Figure 1.10). Each hybrid orbital experiences identical interactions with the other three hybrid orbitals. The four hybrid orbitals must therefore be of equivalent energy: they are degenerate.

Figure 1.9
sp³ hybridisation.

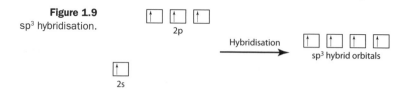

Figure 1.10
Geometry of sp³ hybrid orbitals.

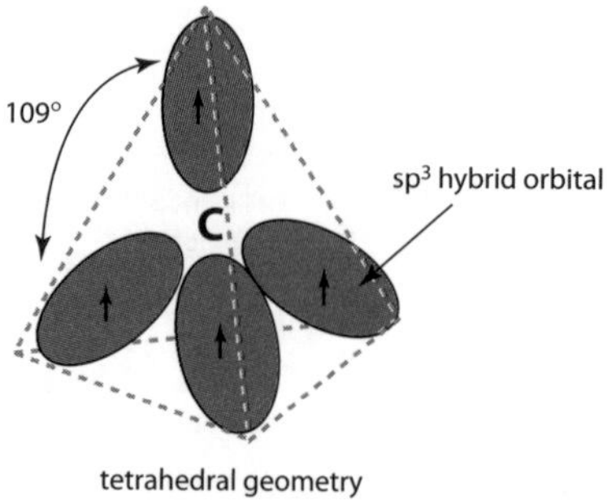

tetrahedral geometry

sp² hybridisation

■ In sp² hybridisation the 2s and two of the 2p orbitals are combined to create three sp² hybrid orbitals. The remaining 2p orbital remains unchanged by the hybridisation process.

■ Each hybrid orbital has a 1/3 contribution from each of the three starting atomic orbitals.

■ The three sp² hybrid orbitals are of an equivalent energy, which is lower than the starting 2p orbitals but higher than the starting 2s orbital.

■ During hybridisation orbitals are not created or destroyed: the three atomic orbitals mix to give three hybrid orbitals.

■ Electrons in hybrid orbitals experience electron–electron repulsion in the same way as electrons in atomic orbitals. Each sp² hybrid orbital will be occupied singly before spin pairing if necessary (Figure 1.11).

■ The three sp² hybrid orbitals adopt a trigonal planar geometry to minimise electron–electron repulsion between the hybrid orbitals (Figure 1.12). Each hybrid orbital experiences identical interactions with the other two hybrid orbitals. The three hybrid orbitals must therefore be of equivalent energy: they are degenerate.

■ The unhybridised 2p orbital will be perpendicular to the plane of the sp² hybrid orbitals as the hybrid orbitals will only have contributions from the p orbitals along two of the coordinate axes, p_x and p_y in Figure 1.12. The remaining p_z orbital is unchanged.

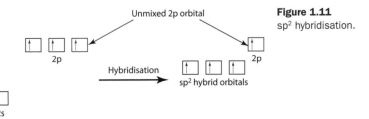

Unmixed 2p orbital

2p

Hybridisation

sp² hybrid orbitals

2p

2s

Figure 1.11
sp² hybridisation.

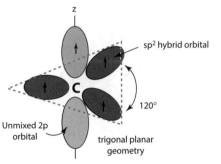

z

sp² hybrid orbital

C

120°

Unmixed 2p orbital

trigonal planar geometry

Figure 1.12
Geometry of sp² hybrid orbitals.

sp hybridisation

- In sp hybridisation the 2s and one of the 2p orbitals are combined to create two sp hybrid orbitals. The two remaining 2p orbitals remain unchanged by the hybridisation process.
- Each hybrid orbital has a 1/2 contribution from each of the two starting atomic orbitals.
- The two sp hybrid orbitals are of an equivalent energy, which is lower than the starting 2p orbital but higher than the starting 2s orbital.
- During hybridisation orbitals are not created or destroyed. The two atomic orbitals mix to give two hybrid orbitals.
- Electrons in hybrid orbitals will experience electron–electron repulsion in the same way as electrons in atomic orbitals. Each sp hybrid orbital will be occupied singly before spin pairing if necessary (Figure 1.13).
- The two sp hybrid orbitals adopt a linear geometry to minimise electron-electron repulsion between the hybrid orbitals (Figure 1.14). Each hybrid orbital experiences an identical interaction with the other hybrid orbital. The two hybrid orbitals must therefore be of equivalent energy: they are degenerate.
- The two unhybridised 2p orbitals will be perpendicular to the sp hybrid orbitals as the hybrid orbitals will only have contributions from the p orbital along one of the coordinate axes, p_x in Figure 1.14. The remaining p_y and p_z orbitals are unchanged.

Figure 1.13
sp hybridisation.

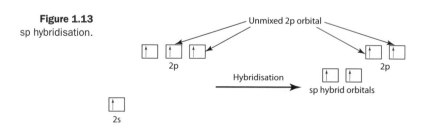

Figure 1.14
Geometry of sp hybrid orbitals.

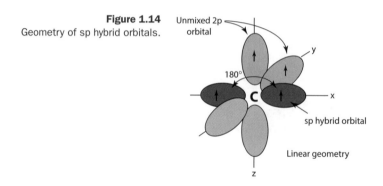

Comparison of hybridisation modes

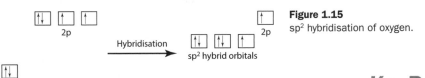

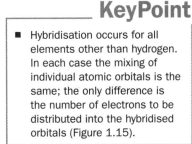

Figure 1.15
sp² hybridisation of oxygen.

- Hybridisation involves the mixing of s and p orbitals to create new orbitals which possess both s and p character.
- The number of p orbitals mixed with the s orbital will affect the shape and character of the hybrid orbital. The more p orbital contribution, the greater p character the new hybrid orbital will have (Figure 1.16).
- sp³ hybrid orbitals have greatest p character, whereas sp hybrid orbitals have greatest s character.

KeyPoint

- Hybridisation occurs for all elements other than hydrogen. In each case the mixing of individual atomic orbitals is the same; the only difference is the number of electrons to be distributed into the hybridised orbitals (Figure 1.15).

Figure 1.16
Comparison of hybridised orbital shape.

sp³   sp²   sp

Bonding

Tip

The two atoms in a double or triple bond must have the same hybridisation – sp² for a double bond and sp for a triple bond.

- σ bonds form from the overlap of two sp³ hybridised orbitals, two sp² hybridised orbitals or two sp hybridised orbitals which each contain a single electron (Figure 1.17). Hydrogen atoms are free to form σ bonds with sp³, sp² and sp hybridised orbitals.

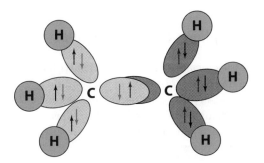

Figure 1.17
Combining hybridised orbitals to form molecules.

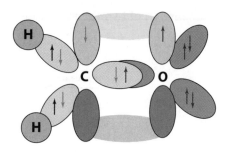

Figure 1.18
π bonding in formaldehyde.

- π bonds form from the overlap of two adjacent p orbitals which each contain a single electron (Figure 1.18).

Valence shell electron pair repulsion (VSEPR)

- When considering bond angles, the repulsion between different types of electron pairs must be considered.
- Lone pair–lone pair repulsion is greater than lone pair–bonding pair repulsion, which in turn is greater than bonding pair–bonding pair repulsion (Figure 1.19).
- In addition to accounting for bond angles in molecules such as ethane, VSEPR accounts for the distortions away from the anticipated hybridisation bond angles, for example the 104.5° bond angle observed in water.

Figure 1.19
Effect of lone pair–lone pair repulsion in water.

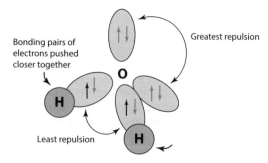

Bonding pairs of electrons pushed closer together

Greatest repulsion

Least repulsion

Tip

There is nothing magical about electronegativity. As you go across the periodic table the number of protons in the nucleus increases but the corresponding electrons are added to the same orbitals. The result is that the electrons are increasingly tightly held as you move from left to right. The 'effective nuclear charge' is increasing. As you go down a period of the periodic table you have additional shells of electrons shielding the nucleus; this reduces the effective nuclear charge.

Nucleophiles and electrophiles

- Molecules that are electron-rich with either a free pair of electrons or a π bond that they can donate are termed nucleophilic (nucleus loving). Since nucleophiles donate a pair of electrons they are also Lewis bases.

- Molecules that are electron-deficient and are able to accept a pair of electrons from a nucleophile are termed electrophilic (electron loving). Since electrophiles accept a pair of electrons they are also Lewis acids.

Drawing chemical structures

- Chemical structures can be represented graphically in a wide range of ways (Figure 1.20).
- A simple molecular formula provides information on the atoms present in a molecule but there is no information concerning connectivity or structure.
- Structural and skeletal formulae both provide additional information concerning the arrangement of atoms within a molecule and their connectivity. Whereas the structural formula displays all atoms present, the skeletal formula focuses on the carbon framework and omits hydrogen atoms attached to carbon, though those attached to other atoms are still included.
- For reaction mechanisms we will adopt the skeletal formula approach.
- Molecular models are useful when considering chemical reactivity and how molecules might interact with each other or with a protein.

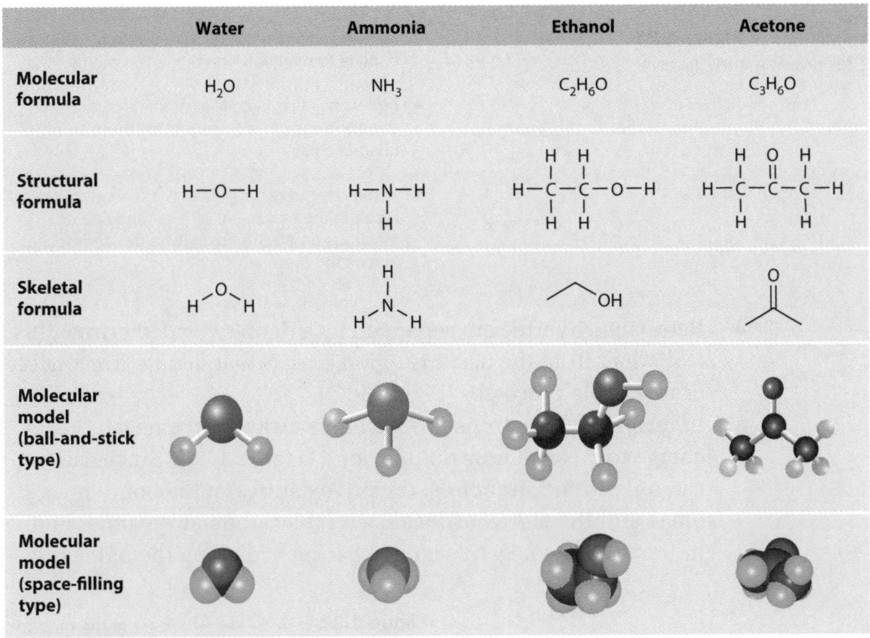

	Water	Ammonia	Ethanol	Acetone
Molecular formula	H_2O	NH_3	C_2H_6O	C_3H_6O
Structural formula	H—O—H	H—N—H $\mid$ H	$\begin{array}{cc} H & H \\ \mid & \mid \\ H-C-C-O-H \\ \mid & \mid \\ H & H \end{array}$	$\begin{array}{ccc} H & O & H \\ \mid & \parallel & \mid \\ H-C-C-C-H \\ \mid & & \mid \\ H & & H \end{array}$
Skeletal formula	H$\diagdown$O$\diagup$H	H$\diagup$N$\diagdown$H $\mid$ H	$\diagup\!\diagdown\!\diagup$OH	(skeletal structure of acetone)
Molecular model (ball-and-stick type)				
Molecular model (space-filling type)				

Figure 1.20 Different ways of representing a molecule.

Tips

Resonance/mesomerism

- The electrons in a π bond are less tightly bound to the individual atoms of the bond than the electrons of a σ bond.
- A conjugated system is a series of overlapping p orbitals which allows for the delocalisation of π electron density across the whole system.
- In a conjugated system an excess or deficiency in electron density associated with one atom can be propagated along the system through the delocalisation of the π bonds.
- This delocalisation of π electron density can be represented pictorially as a series of resonance hybrids or canonical structures. The conjugated system does not exist as a single, individual structure; rather it is a sum of the contributing resonance hybrids.

Drawing resonance hybrids

- Curly or curved arrows ( ) are used to represent the movement of a pair of electrons and to convert between individual resonance hybrids.
- Curly arrows always begin at an electron-rich source and point towards an electron-deficient site (Figure 1.21).

Figure 1.21
Rules for drawing curly arrows.

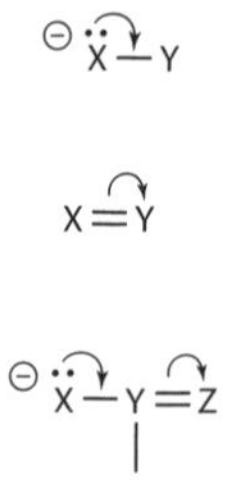

$\ominus \ddot{X} - Y$

- Arrows pointing between atoms indicate that a bond is formed between those two atoms.

$X = Y$

- Arrows pointing directly at atoms indicate that the electrons are located on that atom as an unshared pair.

$\ominus \ddot{X} - Y = Z$

- In the case of resonance one arrow will typically result in the movement of a second pair of electrons as electron density is delocalised across the system.

- Resonance hybrids are separated by a double-headed arrow; this is distinct from the usual reaction arrows that denote irreversible or reversible processes (Figure 1.22).
- In resonance, electrons always move away from a negative charge to create a new double bond (Figure 1.23). Since carbon can only form four bonds, the pre-existing double bond must break and the electrons localise on the carbon at the other end. In this way the negative charge has moved along the molecule.

Figure 1.22
Different types of arrows.

A ⟶ B   Irreversible reaction

A ⇌ B   Reversible reaction

A ⟷ B   Resonance

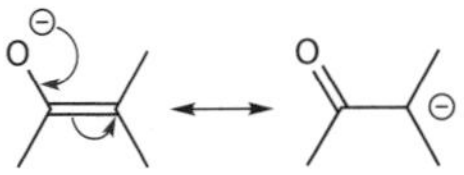

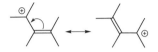

Figure 1.23
Anions in resonance.

■ If there is a positive charge present, the electron pair will move towards this charge through the breaking of one double bond and the creation of a new double bond. In this way it appears as though the positive charge is moving along the molecule, though in reality it is electrons moving in the opposite direction (Figure 1.24).

Tips

■ When moving a negative charge to convert between resonance structures two curly arrows are required.
■ When moving a positive charge to convert between resonance structures only one curly arrow is required.

Figure 1.24
Cations in resonance.

■ Although a conjugated system has the electron pairs delocalised over the entire system, when using the curly arrow notation to convert between resonance hybrids it is necessary to consider the system as a series of localised π bonds.
■ You should *always* draw benzene and its derivatives as localised π bonds (Figure 1.25). The importance of this will be seen later when considering the reactivity and chemistry of benzene and its derivatives.
■ Benzene derivatives experience resonance due to the overlapping of adjacent p orbitals.
■ This resonance can either push electron density into the ring or withdraw electron density from the ring (Figures 1.26 and 1.27).

Figure 1.25
The drawing of benzene.

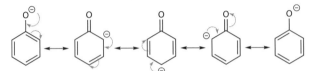

Figure 1.26
Resonance in the phenoxide anion.

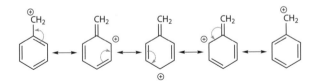

Figure 1.27
Resonance in the benzyl cation.

Self-assessment

1. **Which of the following rules are not required when considering the filling of atomic orbitals?**
a. Aufbau principle
b. valence shell electron pair repulsion
c. Pauli exclusion principle
d. Hund's rule

2. **The following ground state electronic configuration describes which element?**

$$1s^2\ 2s^2\ 2p^5$$

a. nitrogen
b. oxygen
c. fluorine
d. chlorine

3. **Which of the following ground state electronic configurations correctly describes the sodium ion Na^+?**
a. $1s^2\ 2s^2\ 2p^6$
b. $1s^2\ 2s^2\ 2p^6\ 3s^1$
c. $1s^2\ 2s^2\ 2p^5\ 3s^1$
d. $1s^2\ 2s^2\ 2p^5\ 3s^2$

4. **The bent geometry of a water molecule is due to which of the following factors?**
a. The oxygen atom is sp hybridised.
b. Water molecules are able to hydrogen bond to each other.
c. The hydrogen atoms are only weakly bound to the central oxygen atom.
d. Non-bonded pairs of electrons on the oxygen atom are repulsed.

5. **During hybridisation, the sp^2 mode requires the mixing of which orbitals?**
a. one s orbital and three p orbitals
b. one s orbital and two p orbitals
c. one s orbital and one p orbitals
d. only the three p orbitals

6. **Which of the bonds labelled in the following molecule is shortest?**

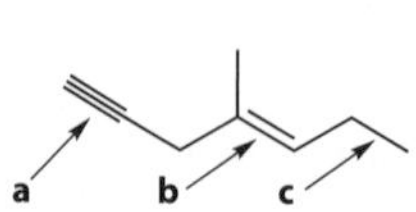

a. bond a
b. bond b
c. bond c
d. they are all carbon–carbon bonds and are therefore the same length

7. **What is the hybridisation of the atom labelled in the following molecule?**

a. sp³ hybridised
b. sp² hybridised
c. sp hybridised
d. both sp² hybridised and sp³ hybridised

8. **Which of the following molecules can undergo resonance?**

a.

b. $\diagdown\diagup\diagdown$ NH₂

c. $\diagup\diagdown\diagup\diagdown$ NH₂

d.   OH

9. **Which of the following phenolate canonical structures is the lowest in energy?**

a.

b.

c.

d.

10. Which of the following functional groups will donate electron density to an adjacent double bond by resonance?

a.   an ester
b.   a nitro group
c.   an amino group
d.   a nitrile group

chapter 2
Intermolecular interactions

Overview

After learning the material presented in this chapter you should:

- know the different types of intermolecular forces involved in drug–water, drug–membrane and drug–receptor interactions
- appreciate the relative strengths of the different intermolecular interactions
- understand the hydrophobic effect and its significance as regards drug–receptor interactions, protein folding and surfactant micelle formation
- be able to use empirical formulae to calculate the potential energies of van der Waals and electrostatic interactions.

■ Intermolecular interactions describe the *non-covalent* interactions that exist between molecules.

■ The forces involved in intermolecular interactions are much weaker than those involved in the ionic bonds in substances like sodium chloride, and they are weaker than the forces involved in the sigma bonds (σ bonds) and pi bonds (π bonds) which link together atoms in covalent compounds.

■ The different types of intermolecular interactions include: *dipole–dipole* and *van der Waals* interactions, *Coulombic* (or *electrostatic*) interactions and *hydrogen bonds*.

■ Each of these various interactions contributes to the *enthalpy* of an intermolecular system.

■ The enthalpy of a system is a measure of the total energy in that system, and is measured using the SI unit of joules. The total enthalpy of a system cannot be measured directly, and so when describing the process of two molecules interacting with one another, it is usual to refer only to the *change in enthalpy* (ΔH) that results as a consequence of the interaction between the molecules. If ΔH is negative, the interaction between the molecules leads to a release of heat and the process is said to be *exothermic*. If ΔH is positive, the interaction between the molecules requires input of energy, and the process is said to be *endothermic*.

■ The relative strengths of the different intermolecular forces are given in Table 2.1.

■ There is invariably also an *entropic* contribution to the stability of intermolecular interactions, which arises because of a change in the degree of disorder in the system before and after the intermolecular interactions are established. When two molecules are bound together via non-covalent interactions, their freedom of

Table 2.1 Strengths of intermolecular and intramolecular forces

Interaction/force	Example	Potential energy (kJ.mol^{-1})
Covalent bond	C–C single bond	391
Ionic bond	NaCl lattice energy	−787
Coulombic/electrostatic interaction	Drug NH_3^+ and protein COO^- at 4 Å separation in receptor-binding site	87
Hydrogen bond	C=O....H–N hydrogen bond with O...N ~3 Å	30
Dipole–dipole interaction	H_2O ... H_2O	~ 2
van der Waals interaction (London dispersion force)	Two CH_3 groups in contact with one another	< 1

movement and their freedom to change conformation become more restricted, and such changes work to reduce the entropy of the system.

- The *hydrophobic effect* describes an entropic contribution to the stability of an intermolecular interaction that arises because of changes in the behaviour of the solvent (water) in the system. This effect is significant for protein folding, and in the formation of surfactant micelles in water.

  - In *protein folding*, the removal of the non-polar amino acid residues to the interior of the protein molecule – away from contact with the surrounding water molecules – results in water molecules being liberated to the bulk (to behave as liquid water), and this leads to a significant increase in the water entropy.
  - In *surfactant micelle formation*, it is the removal of the non-polar hydrocarbon chain(s) of the surfactant molecules to the micelle interior – away from contact with water – that leads to water molecules being liberated to the bulk, and thence to an increase in solvent entropy.

KeyPoints

- Intermolecular interactions are non-covalent interactions.
- The forces involved in intermolecular interactions are weaker than those involved in ionic or covalent bonds.
- Intermolecular interactions include van der Waals (dipole–dipole and dipole–induced dipole) interactions, hydrogen bonds and electrostatic interactions.

Importance of intermolecular interactions in pharmacy

It is the nature and the extent of the intermolecular interactions that exist between a drug and its target receptor that determine the efficacy of the drug.

It is the relative strengths of the intermolecular interactions that exist between a drug and its surrounding solvent, and between a drug and the lipid molecules that make up a cell membrane, that determine how readily the drug passes into and across the membrane and how readily, therefore, it is absorbed into cells.

It is by exploiting the intermolecular interactions between drugs and surfactant molecules that the various colloidal drug delivery vehicles like micelles and microemulsions are produced.

Coulombic/electrostatic interactions

Drug molecules that carry positive- or negative-charged groups will tend to interact more favourably with water, by virtue of the electrostatic/Coulombic forces between their charged groups and the water dipoles (Figure 2.1).

When drugs with positive- and/or negative-charged groups bind in the binding site at their target receptor, the complex formed will be stabilised significantly by any *ionic interactions* between the charged groups on the drug and oppositely charged groups located in complementary positions within the binding site.

The favourable interactions formed between charged drug molecules and their surrounding solvent mean that charged drug molecules do not readily cross cell membranes – there is an energy penalty associated with the stripping of the hydration shell from the charged groups (otherwise referred to as *desolvation* of the charged groups).

The strength of an electrostatic interaction can be quantified by the associated potential energy of the interaction (U_c), which is computed according to the empirical formula:

$$U_c = 332 \cdot \frac{q_1 \cdot q_2}{\varepsilon \cdot r_{12}}$$

where q_1 and q_2 are the charges on the two charged groups involved in the interaction (measured in electron charges), r_{12} is their distance

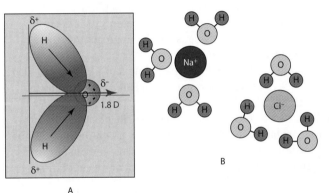

A

B

Figure 2.1

Water dipole and ion–dipole electrostatic interactions.

(A) Water dipole ($\mu = 1.85$ Debye) – arising because of the uneven sharing of electrons between its oxygen and hydrogen atoms.

(B) Ion–dipole electrostatic interactions between the formally charged Na^+ and Cl^- ions and their surrounding water dipoles.

of separation (in Å), and epsilon (ε) is the bulk dielectric constant of their surrounding/intervening medium. The dielectric constant for water (at 298 K) is ~80, and the dielectric constant for the interior (so-called *hydrophobic core*) of a globular protein is ~4.

The dielectric constant for a given medium, ε, is the ratio of the permittivity in that medium relative to that in a vacuum; it thus has no units.

van der Waals interactions

van der Waals forces are the weakest of the intermolecular forces. They include the forces that arise through *dipole–dipole, dipole–induced dipole* and *transient dipole–induced dipole* interactions.

In molecules where there are covalent bonds joining atoms that have a (marked) difference in *electronegativity*, the atoms involved will not share the bond electrons equally, with the result that the bond is *polarised*. Each of the joined atoms in this case carries a *partial charge* – one of the atoms carries a partial positive charge and the other a partial negative charge. Molecules of this type are said to have a *permanent dipole* (Figure 2.1).

Molecules with permanent dipoles interact with one another, and also with molecules that carry full electrical charges (i.e. molecular ions), via electrostatic forces.

The resultant of all dipoles within a molecule is quantified by its *dipole moment* μ (which is measured in Debye). Dimethyl sulfoxide (DMSO), pyridine and indole have dipole moments of 3.96 D, 2.03 D and 2.11 D, respectively.

In drug molecules, there are permanent dipoles associated with any carbon–oxygen, carbon–halogen and carbon–nitrogen bonds, and the common functionalities involved include hydroxyl groups and amine groups.

A force of attraction also arises when a molecule with a permanent dipole approaches close to a molecule that lacks a permanent dipole, but which, because of the approaching dipolar molecule, develops an *induced dipole*.

Two molecules, both of which lack permanent dipoles, can also experience a force of attraction, due to *transient dipole–induced dipole* interactions. All molecules (even those lacking permanent dipoles) will have a transient dipole, because the electrons in the molecule are constantly moving and so, at any given instant, will be unevenly distributed across covalent bonds. This uneven electron distribution in one molecule leads to the momentary generation of

> **Tip**
>
> Water has an anomalously high melting point and boiling point because of its large dipole moment (1.85 Debye), and because of its extensive network of hydrogen bonds involving the oxygen lone pair electrons and the protons on neighbouring water molecules.

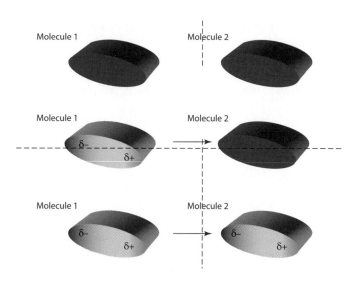

Figure 2.2
Transient dipole–induced dipole interactions.

a polarisation within a nearby molecule, and the result is a system with interacting dipoles (Figure 2.2).

van der Waals attraction is at a maximum when the interacting molecules are in intimate contact, with the contacting atoms separated by a distance which corresponds to the sum of their *van der Waals radii* (Figure 2.3).

Molecules 1 and 2 have no net dipole (Figure 2.2, top). At a given instant, molecule 1 develops a transient dipole and approaches molecule 2 (Figure 2.2, centre); as it does so, its transient dipole causes an induced dipole to be set up within molecule 2 (Figure 2.2, bottom).

Molecules whose atoms are forced to approach one another closer than the sum of their van der Waals radii will experience a

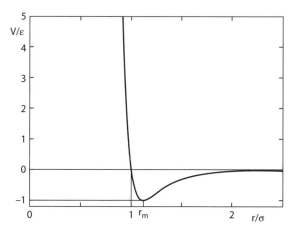

Figure 2.3
Lennard-Jones potential.

repulsive force, which increases very sharply the closer the atoms are forced together (Figure 2.3).

The strength of a van der Waals interaction can be quantified by the associated potential energy of the interaction (U_{VDW}), which is computed according to the empirical (Lennard-Jones) potential:

$$U_{VDW} = 4\varepsilon \cdot \left[\left(\frac{\sigma}{r_{12}} \right)^{12} - \left(\frac{\sigma}{r_{12}} \right)^{6} \right]$$

where σ is the distance at which the interaction energy is zero (in Å), ε measures the maximum strength of interaction for the interacting atoms (in J.mol^{-1}) and r_{12} is their distance of separation (in Å).

The Lennard-Jones interaction potential energy (V) is a function of the distance of atomic separation, r (the former normalised with respect to ε and the latter with respect to σ). r_m is the distance at which the potential energy is a minimum, and σ is the interatomic separation when the energy of interaction is zero (Figure 2.3).

KeyPoint

- van der Waals interactions are maximised when molecules are in intimate contact with one another (that is, when they have atoms separated by a distance equal to the sum of their van der Waals radii.

Hydrogen bond interactions

Hydrogen bonds are attractive interactions formed between *hydrogen bond donor* and *hydrogen bond acceptor* groups.

Hydrogen bond donor groups are chemical groups that have a hydrogen atom bonded to an *electronegative* atom (in drug molecules, generally, O and N atoms). Hydrogen bond donor groups commonly found in drug molecules include hydroxyl (OH) groups, amides (CONH) and primary and secondary amine (RNH$_2$ and R$_2$NH) groups.

Hydrogen bond acceptor groups are chemical groups that have an atom with one or more lone pairs of electrons. Hydrogen bond acceptor groups commonly found in drug molecules include fluoro- and chloro- groups, carbonyl (C=O) groups, amides (CONH), and primary, secondary and tertiary amine (NH$_2$, R$_2$NH and R$_3$N) groups.

The hydrogen atom involved in a hydrogen bond is effectively 'shared' between the two heteroatoms involved in the acceptor and donor groups.

The force of (electrostatic) attraction associated with the hydrogen bond is sufficiently strong that the acceptor and donor heteroatoms approach closer than the sum of

KeyPoint

- Hydrogen bonds involve the sharing of a proton between two electronegative atoms (typically, O or N).

their van der Waals radii. The hydrogen bond formed between the peptide CO and NH groups within the alpha helix of a folded protein, for example, typically has the O and N atoms at around 2.8 Å separation, whereas the sum of the van der Waals radii for the O and N atoms is 1.52 + 1.55 = 3.07 Å.

Molecular affinity and the free energy of interaction

Interactions between molecules can involve any or all of the various different intermolecular forces outlined above.

The change in Gibbs free energy that is associated with an interaction between two molecules (ΔG, J.mol^{-1}) is (1) a function of the change in enthalpy in the system (ΔH, J.mol^{-1}), which arises because of changes in the molecules' bonding arrangements (e.g. the formation of hydrogen bonds between the interacting molecules) and (2) because of changes in the system entropy (ΔS), which stem from changes in the molecules' orientational and conformational freedom.

The equation that relates ΔG, ΔH and ΔS is simply:

$$\Delta G = \Delta H - T \cdot \Delta S$$

where T is the temperature (in Kelvin).

The change in Gibbs free energy associated with any intermolecular interaction provides a measure of the *affinity* between the two molecules. The lower (the more negative) the associated ΔG, the greater the affinity of the interacting molecules for one another.

> **Tip**
>
> The oral availability of the majority of drug molecules falls within the criteria of *Lipinski's rule of five*: molecules that are orally active will generally have a molecular mass < 500, a log(octanol–water partition coefficient) no greater than 5, contain no more than five hydrogen bond donor groups and have no more than 10 hydrogen bond acceptor groups.

Self-assessment

1. **The strength of interatomic interactions follows the order:**
a. ionic > dipole–dipole > van der Waals
b. hydrogen bond > van der Waals > dipole–dipole
c. covalent bond > van der Waals > hydrogen bond
d. van der Waals > hydrogen bond > dipole–dipole

2. **If there is a net gain in entropy associated with an intermolecular interaction, and if the associated enthalpy gain is twice the gain in entropy, the change in Gibbs free energy for the interaction will be:**
a. $\Delta G = 3\Delta S$
b. $\Delta G = \Delta S$
c. $\Delta G = (2 - T)\,\Delta S$
d. $\Delta G = (T - 2)\,\Delta S$

3. **How many hydrogen bond donor groups are found in the analgesic drug paracetamol (below, right)?**

   a. 1
   b. 2
   c. 3
   d. 4

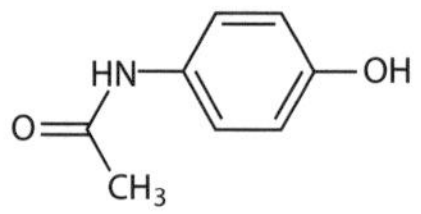

4. **The general anaesthetic, halothane (below right) will *not* be attracted to other molecules of halothane as a result of:**

   a. hydrogen bond interactions
   b. ionic interactions
   c. van der Waals forces
   d. dipole–dipole interactions

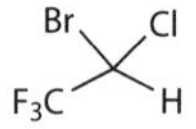

5. **Given a near-perfect complementarity between aspirin (below) and the binding site on its target enzyme, cyclooxygenase, which of the following interactions will contribute *most* to the stability of the system when the drug inserts in the enzyme-binding site?**

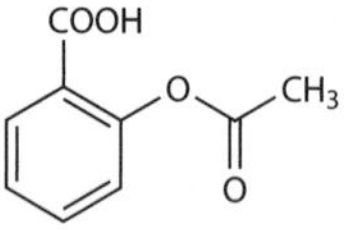

   a. van der Waals interactions involving the drug's methyl group
   b. the hydrogen bond formed by the drug's carbonyl group
   c. the ionic interaction(s) involving the drug's ionised carboxyl group
   d. dipole–dipole interactions involving the drug's ester group

6. **The antibiotic amikacin (below) is predicted to have low oral bioavailability and this can be explained by the fact that:**

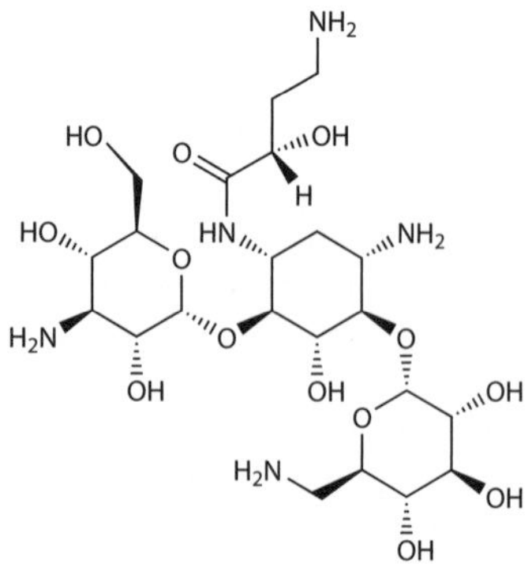

   a. it has too many rings
   b. it has a molecular mass < 500
   c. it has no charged groups
   d. it has too many hydrogen bond acceptor groups

7. **A cationic drug molecule is *unlikely* to bind and become buried within a protein-binding site if the binding site features a prominent:**
a. phosphate group
b. primary amine group
c. sulfonate group
d. carboxylate group

8, **Molecules whose constituent atoms exhibit little variation in electronegativity:**
a. will have high dipole moments
b. will not experience a significant force of mutual attraction as a result of van der Waals interactions
c. are unlikely to contain hydrogen bond donor groups
d. will interact with one another primarily through ionic interactions

9. **Drug molecules based on open chain structures, when compared against those that have the same pattern of functional groups but are based around aromatic and fused ring systems:**
a. will tend to have a greater affinity for their target receptor because their conformational entropy will increase as a result of complex formation
b. will tend to have lower affinity for their target receptor because their conformational entropy will decrease as a result of complex formation
c. will tend to have greater affinity for their target receptor because their conformational entropy will decrease as a result of complex formation
d. will tend to have lower affinity for their target receptor because their conformational entropy will increase as a result of complex formation

10. **If two drugs, A and B, bind in an enzyme's active site and drug A occupies all of the available space while drug B occupies only some of the space:**
a. drug A is likely to have a greater affinity for the enzyme compared with drug B because of increased van der Waals interactions
b. drug B is likely to have a greater affinity for the enzyme compared with drug A because of increased van der Waals interactions
c. drug A is likely to have a greater affinity for the enzyme compared with drug B because of the reduction in solvent entropy caused through water elimination from the active site
d. drug B is likely to have a greater affinity for the enzyme compared with drug A because of the reduction in solvent entropy caused through water elimination from the active site

chapter 3

Acids and bases

Overview

After learning the material presented in this chapter you should:
- be able to identify acidic and basic functional groups in drug molecules
- be able to assess the strength of an acid or base
- know the pK_a values of common acidic and basic groups
- understand why knowledge of acids and bases is important in pharmacy
- be able to use the pK_a values of acidic and basic groups to determine the ionisation state of a drug molecule at a specific pH

Definitions of acids and bases

The Brønsted–Lowry definition of an acid is a species able to donate a proton (H^+), and the corresponding definition for a base is a species able to accept a proton.

The (negatively charged/anionic) species that remains after an acid has lost a proton is known as the *conjugate base*, and the (positively charged/cationic) species created when a base accepts a proton is the *conjugate acid*.

The reaction between an acid and base can thus be written as the word equation:

$$acid + base \rightleftharpoons conjugate\ acid + conjugate\ base$$

and if we consider the reaction between an acid *HA* and a base *B* we have the equation:

$$HA + B \rightleftharpoons A^- + HB^+$$

where A^- is the conjugate base and HB^+ is the conjugate acid.

Acid and base strength

The strength of an acid describes how readily it yields a proton when in solution. A *strong acid* is one that is fully ionised (that is, fully dissociated) in (aqueous) solution. Examples of strong inorganic acids include hydrochloric acid (HCl), sulfuric acid (H_2SO_4) and nitric acid (HNO_3). When hydrogen chloride (gas) is dissolved in water it dissociates completely to produce proton and chloride ions:

$$HCl \rightleftharpoons H^+ + Cl^-$$

A *weak acid* is only partially dissociated in solution. Examples of weak acids include carbonic acid (H_2CO_3) and ethanoic acid (also known as acetic acid, CH_3COOH). The solution of a weak acid contains both the free (unionised) acid

Tip

The names of drug salts can often be used to tell whether the parent drug is acidic or basic. So, for example, we can deduce that the opiate morphine must be *basic* because its commonly used salts, morphine *hydrochloride* and morphine *sulfate*, indicate that they are produced by the addition of a weak base (morphine) to a strong acid – for these two salts, either *hydrochloric* acid or *sulfuric* acid. Likewise, the name of the antibiotic salt, penicillin G sodium, indicates that it is formed by the addition of a weak acid (penicillin G, also known as the benzyl penicillin) to a strong base; the strong base in this case is *sodium* hydroxide.

and the conjugate base. In solution, therefore, ethanoic acid exists in equilibrium with ethanoate (CH_3COO^-) ions:

$$CH_3COOH \rightleftharpoons CH_3COO^- + H^+$$

In like manner, a *strong base* is a base that is fully ionised in aqueous solution and a *weak base* is one that is only partially ionised in aqueous solution. Examples of strong bases include sodium hydroxide (NaOH) and potassium hydroxide (KOH). In solution, sodium hydroxide is fully dissociated as sodium and hydroxide ions:

$$NaOH \rightarrow Na^+ + OH^-$$

Examples of *weak bases* include ammonia (NH_3) and methylamine (CH_3NH_2). An aqueous solution of ammonia (gas) contains an equilibrium mixture of the free base (ammonia) and the conjugate acid (an ammonium ion, NH_4^+):

$$NH_3 + H_2O \rightleftharpoons NH_4^+ + OH^-$$

Quantifying the strengths of acids and bases: pK_a and pK_b values

The strength of an acid in aqueous solution is quantified by its *acid dissocation constant*, K_a. If we consider the equilibrium that exists for an acid *HA* in solution:

$$HA \rightleftharpoons A^- + H^+$$

KeyPoints

- Strong acids are fully dissociated in aqueous solution.
- Weak acids are only partially dissociated in aqueous solution.
- Strong bases are fully ionised in aqueous solution.
- Weak bases are only partially ionised in aqueous solution.
- The strength of an acid or base can be quantified by its pK_a.

K_a is obtained as:

$$K_a = \frac{[A^-] \cdot [H^+]}{[HA]}$$

where [...] indicate the molar concentrations of each of the species, *HA*, A^- and H^+.

The stronger the acid, the more readily it dissociates in solution (forming A^- and H^+ ions) and so the higher its K_a value.

The negative logarithm of K_a ($-\log_{10}K_a$) is the pK_a of the acid (p$K_a = \log_{10}(1/K_a)$). Strong acids have high K_a values and so have low pK_a values. Weak acids have low K_a values and so have high pK_a values.

The strength of a base in solution can be quantified by its *association constant*, K_b. If we consider the equilibrium that exists for a base B in aqueous solution:

$$B + H_2O \rightleftharpoons OH^- + HB^+$$

K_b is obtained as:

$$K_b = \frac{\left[OH^-\right]\cdot\left[HB^+\right]}{[B]}$$

where [...] indicate the molar concentrations of each of the species, B, OH^- and HB^+.

The stronger the base, the more readily it is protonated in solution (forming HB^+) and so the higher its K_b value.

The negative logarithm of K_b ($-\log_{10}K_b$) is the pK_b of the base. Strong bases have high K_b values and so have low pK_b values. Weak bases have low K_b values and so have high pK_b values.

The strength of a base can also be quantified by means of the pK_a of its conjugate acid. The pK_a value for a base is related to its pK_b value as:

$$pK_b = pK_w - pK_a$$

where K_w is the ionic product for water:

$$K_w = \left[OH^-\right]\cdot\left[H^+\right]$$

In aqueous solutions at 25°C, $K_w \sim 10^{-14}$ M, and so $pK_w \sim 14$ and:

$$pK_b = 14 - pK_a$$

The percentage ionisation for an acidic drug can be calculated as:

$$\text{Per cent ionisation} = \frac{100}{1 + \left(\text{antilog}\left(pK_a - pH\right)\right)}$$

and for a basic drug it is calculated as:

$$\text{Per cent ionisation} = \frac{100}{1 + \left(\text{antilog}\left(pH - pK_a\right)\right)}$$

Importance of acids and bases in pharmacy

The majority of drugs contain weakly acidic or weakly basic functional groups. There are

Tip

Solubility for acidic and basic drugs will vary with the pH of their solution. An acidic drug will be most soluble under alkaline conditions and will be least soluble under acidic conditions. Basic drugs will have highest solubility in acidic conditions and least solubility in alkaline conditions. An amphoteric drug will have its lowest solubility at its isoelectric point.

Tips

- Weakly acidic drugs are essentially totally ionised at pH values ≥ 2 units above their pK_a. They are almost totally unionised at pH values ≥ 2 units below their pK_a and are 50% ionised at pH values equal to their pK_a.
- Weakly basic drugs are essentially totally ionised at pH values ≥ 2 units below their pK_a. They are essentially totally unionised at pH values ≥ 2 units above their pK_a and are 50% ionised at pH values equal to their pK_a.

many too that contain both types of group and such compounds are described as *amphoteric* compounds.

The ionisation state of a drug that contains acidic and/or basic functional groups will depend on the pH of the medium it is dissolved in. This in turn will influence the drug's solubility, the way in which it is formulated as a medicine, its absorption and distribution within the body and its pharmacological activity.

Many of the pharmaceutical excipients that are used in formulating drugs as medicines also contain acidic and/or basic functional groups.

Common acidic and basic functional groups

The chemical structures and pK_a values for various acidic and basic functional groups that are commonly found in drug substances are shown in Table 3.1, together with examples of drugs in which they are found. Table 3.2 provides a similar listing for some of the commonly used pharmaceutical excipients that contain acidic and/or basic functional groups.

Table 3.1 Acidic and basic functional groups commonly found in drug substances

Acidic groups		
Carboxyl	R-COOH	Ibuprofen (analgesic)
(pK_a 2–4)		
Phenol	Ar—OH	Phenol (used in antiseptic mouth spray)
(pK_a ~10)		
Sulfonamide	$R-SO_2NH_2$	Diclofenamide (antiepileptic)
(pK_a 8–10)		
Imide	R-CO-NH-CO-R	Phenytoin (antiepileptic)
(pK_a 8–10)		

Basic groups

Primary aliphatic amine	R-NH$_2$	Amfetamine (for treatment of attention deficit hyperactivity disorder)
(pK$_a$ 9–10)		

Secondary aliphatic amine	R-NH-R	Isoprenaline (β$_2$-agonist)
(pK$_a$ 9–10)		

Tertiary aliphatic amine	R-N(R)$_2$	Amitriptyline (antidepressant)
(pK$_a$ 9–10)		

Quaternary amine (pK$_a$ n/a; cationic species exists at all pHs)	R-N$^+$(R)$_3$	Bethanechol (for treating postsurgical urine retention)

Aromatic amine (pK$_a$ 2–4)	Ar—NH$_2$	Benzocaine (local anaesthetic)

Pyridine (pK$_a$ ~5.5)		Pyridoxine (vitamin B$_6$)

Imidazole (pK$_a$ 6–7)		Dacarbazine (anticancer agent)

Table 3.2 Acidic and basic pharmaceutical excipients

Stearic acid (tablet lubricant)	
Aspartame (sweetener)	
Sodium benzoate (preservative)	
Citric acid (soluble tablet disintegrant)	
Butylparaben (antimicrobial preservative)	
Glycine (buffering agent and lyophilisation aid)	

pH and pK$_a$

The pH of a solution is the negative logarithm of its hydrogen ion concentration, that is, $pH = -\log_{10}[H^+]$, or, alternatively, $pH = 1/\log_{10}[H^+]$.

For an aqueous solution of a weak acid, HA, which has an acid dissociation constant, K_a:

$$K_a = \frac{[A^-] \cdot [H^+]}{[HA]}$$

The pH of the solution and the acid pK_a are related by the Henderson–Hasselbalch equation:

$$pH = pK_a + \log \frac{[A^-]}{[HA]}$$

where […] indicate the molar concentrations of each of the species, HA and A^-.

Amphoteric compounds

Substances that contain both acidic and basic functional groups are known as *amphoteric* substances. An amphoteric molecule that has both its acidic and basic groups ionised simultaneously is known as *zwitterionic*.

Amphoteric compounds can potentially exist as cationic, anionic, zwitterionic or neutral species in aqueous solution and their precise ionisation state varies according to the pH of the solution and the pK_a values of their acidic and basic groups.

An amphoteric molecule that has one acidic group and one basic group, with pK_a values of pK_a^1 and pK_a^2, will exist as electrically neutral species when the solution pH is midway between pK_a^1 and pK_a^2, that is:

$$\mathrm{pH} = \frac{pK_a^1 + pK_a^2}{2}$$

Simple buffer solutions

A *buffer* solution is a solution that resists (major) changes in pH. An *acidic buffer* is one that has a pH below 7. An *alkaline buffer* is one that has a pH above 7.

Acidic buffers can be produced simply as a solution of a weak acid and one of its salts. The composition of an acidic buffer and its resulting pH are related according to the equation:

$$K_a = \frac{[\text{salt}] \cdot [\text{H}^+]}{[\text{acid}]}$$

where […] indicate the molar concentrations of hydrogen ions, the acid and its salt.

Alkaline buffers can be prepared simply by preparing a solution containing a weak base and one of its salts. The composition of an alkaline buffer and its resulting pH are related according to the equation:

$$K_a = \frac{[\text{base}] \cdot [\text{H}^+]}{[\text{salt}]}$$

where […] indicate the molar concentrations of hydrogen ions, the acid and its salt.

Simple acidic and alkaline buffers are generally effective in limiting changes in pH over a pH range within 1 log unit of the pK_a of the constituent acid or base. So, for example, an acidic buffer based on benzoic acid, which has a pK_a of 4.2, will only buffer effectively over the pH range 4.2 ± 1.0, that is from pH 3.2 to pH 5.2.

Universal buffer solutions

A *universal buffer* solution is a buffer solution that buffers over a wide pH range. Such buffers can be prepared using substances that have two or more functional groups with pK_a values separated by ≥2 pH units. Citric acid, for example, has three acidic groups, with pK_a values of 3.06, 4.78 and 5.40, and so it can buffer over the pH range 2.1–6.4.

Universal buffers that can buffer over even broader pH ranges are prepared using mixtures of substances whose pK_a values are separated by ≥2 pH units. A mixture of citric acid, boric acid, diethyl barbituric acid and potassium dihydrogen phosphate provides a universal buffer with a working pH range of 2.4–12.

Self-assessment

1. In aqueous solution at pH 7, the molecules of the appetite suppressant phentermine (shown below) will exist predominantly as:
    a. cations
    b. anions
    c. neutral molecules
    d. unionised species

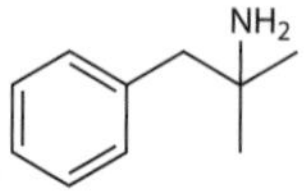

2. In an aqueous solution pH 4.0, the concentration of hydrogen ions will be:
    a. 10^4 mM
    b. 10^{-4} M
    c. 10^4 M
    d. 10^{-4} mM

3. Aspirin (shown below) has a functional group with a pK_a value of:
    a. 0.5
    b. 10.5
    c. 7.5
    d. 3.5

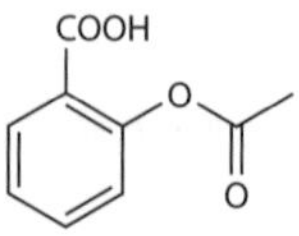

4. Ephedrine hydrochloride is used in the treatment of asthma. Ephedrine is:
    a. a weak acid
    b. a strong acid
    c. a weak base
    d. a strong base

5. Glycine (shown below) has pK_a values of 2.34 and 9.60. Its isoelectric point will be: 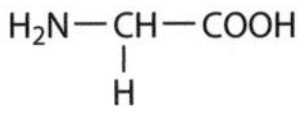
a. 11.94
b. 7.26

$$H_2N-CH-COOH$$
$$|$$
$$H$$

c. 5.97
d. 7.00

6. In aqueous solution pH 9, the motion sickness drug diphenhydramine (shown below), is likely to exist:

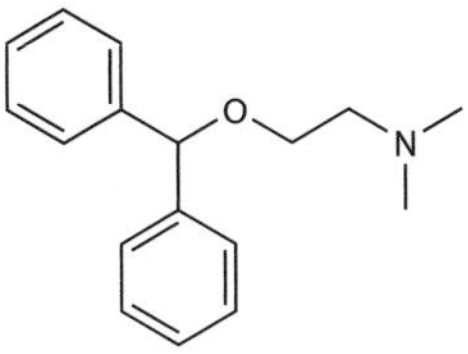

a. predominantly as uncharged molecules
b. as a 50:50 mix of uncharged and negatively charged molecules
c. as a 50:50 mix of uncharged and positively charged molecules
d. predominantly as positively charged molecules

7. Given that ethanoic acid has a pK_a of 4.76, a buffer solution that comprises an equimolar mixture of sodium ethanoate and ethanoic acid will have a working pH in the range:
a. 2.76–6.76
b. 0–4.76
c. 4.7–6.76
d. 3.76–5.76

8. You need to prepare a solution buffered in the range 8.6–10.6. Given that the pK_a of the ammonium ion is 9.25, what is the approximate ratio of ammonia and ammonium chloride that you will require?
a. 2:1 (ammonia : ammonium chloride)
b. 1:2 (ammonia : ammonium chloride)
c. 1:1 (ammonia : ammonium chloride)
d. 2:2 (ammonia : ammonium chloride)

9. At pH 7.4, the local anaesthetic benzocaine will be:

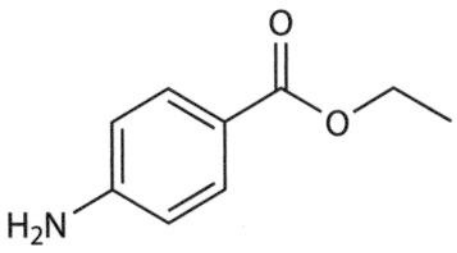

a. neutral
b. anionic
c. cationic
d. zwitterionic

10. Which of the drugs shown below is amphoteric?

a.

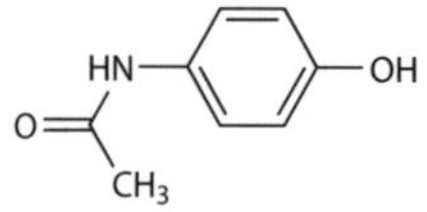

b.

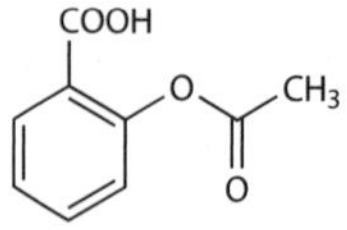

c.

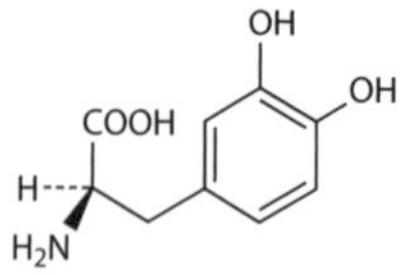

d.

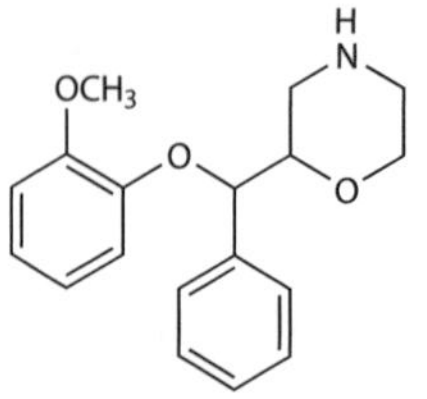

chapter 4
Stereochemistry

Overview

After learning the material presented in this chapter you should:

- understand the difference between structural isomers and stereoisomers
- be able to determine the possible positional isomers and functional group isomers for a given molecular formula
- be able to recognise chiral carbons and draw enantiomers
- be able to recognise diastereomers and the meso-forms of molecules
- understand optical activity and the meaning of the terms levorotatory, dextrorotatory and racemic mixture
- know about the Fischer (L- and D-) system of stereochemical nomenclature
- be able to determine the stereochemistry of a molecule using the Cahn–Ingold–Prelog (R- and S-) nomenclature
- understand the conformational isomerism of open-chain molecules and the meaning of gauche, anti, staggered and eclipsed bonds
- understand the conformational isomerism of cyclic molecules and recognise chair, boat and half-chair conformation
- understand the configurational/geometric isomerism of molecules with double bonds and be able to identify *cis*- and *trans*- and *E*- and *Z*-isomers
- understand topological isomerism in reference to biological macromolecules.

Stereochemistry and its importance in pharmaceutical science

Stereochemistry is concerned with the spatial arrangements of atoms in molecules, and the effects that these have on the physical and chemical properties of the molecules.

A detailed understanding of stereochemistry is vital in pharmaceutical science:

- The clinical efficacy of a drug is critically dependent on its three-dimensional structure.
- The stereochemistry of a drug can impact upon its toxicity and side effects.

Molecules that have the same molecular formula but different chemical structures are known as *isomers*.

Molecules that have the same molecular formula but different patterns of covalent bonding are known as *constitutional isomers* or *structural isomers*. Structural isomers have different International Union of Pure and Applied Chemistry (IUPAC) names.

The hydrocarbon butane (molecular formula, C_4H_{10}) has two structural isomers: *n*-butane and isobutane (Figure 4.1).

There are three structural isomers with molecular formula C_3H_8O: propan-1-ol, propan-2-ol and methoxyethane (Figure 4.1). Structural isomers that have the same

Figure 4.1
Structural isomers.

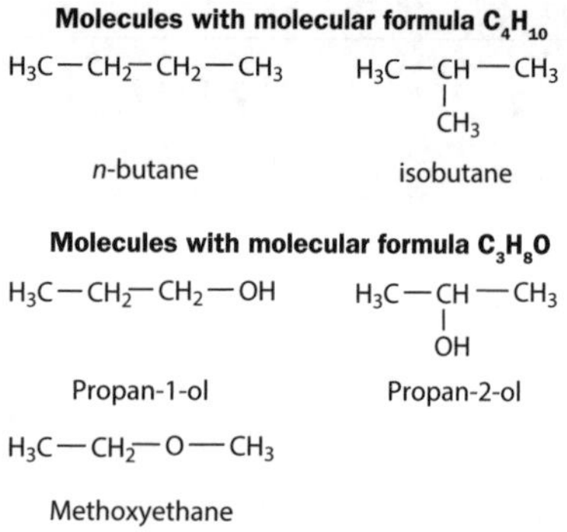

Molecules with molecular formula C_4H_{10}

$H_3C-CH_2-CH_2-CH_3$

$H_3C-CH-CH_3$
$\quad\quad\quad | $
$\quad\quad\quad CH_3$

n-butane

isobutane

Molecules with molecular formula C_3H_8O

$H_3C-CH_2-CH_2-OH$

$H_3C-CH-CH_3$
$\quad\quad\quad | $
$\quad\quad\quad OH$

Propan-1-ol

Propan-2-ol

$H_3C-CH_2-O-CH_3$

Methoxyethane

KeyPoints

- Structural isomers have the same molecular formula but different patterns of covalent bonding.
- Stereoisomers have the same patterns of covalent bonding but different spatial arrangements of their atoms/ functional groups.

functional group attached at different positions (like propan-1-ol and propan-2-ol) are known as *positional isomers*. Structural isomers that have different functional groups (like propan-1-ol and methoxyethane) are known as *functional group isomers*.

Stereoisomers

- Molecules that have the same pattern of covalent bonding but different spatial arrangements of their atoms and functional groups are known as *stereoisomers*.
- Stereoisomers that are non-superimposable mirror images are known as *enantiomers*.
- Molecules that exist as enantiomers are sometimes referred to as *chiral* compounds (and are said to show *chirality*, that is, *handedness*) because they always contain at least one *chiral centre* – usually a tetrahedral carbon atom (sometimes also referred to as a stereogenic centre or an *asymmetric carbon*). Molecules that lack chiral centres are referred to as *achiral* compounds.
- Chiral carbons are tetrahedral (sp³ hybridised) carbon atoms that have four different chemical groups attached (Figure 4.2).
- Enantiomers have the same chemical properties as one another, except that they sometimes show different rates of chemical reaction with other chiral compounds.
- Enantiomers have the same physical properties as one another, except in relation to their *optical activity*, that is, the way in which they rotate the plane of *plane-polarised light* (one enantiomer will rotate the plane in one direction and the other will rotate it by an identical amount in the opposite direction).

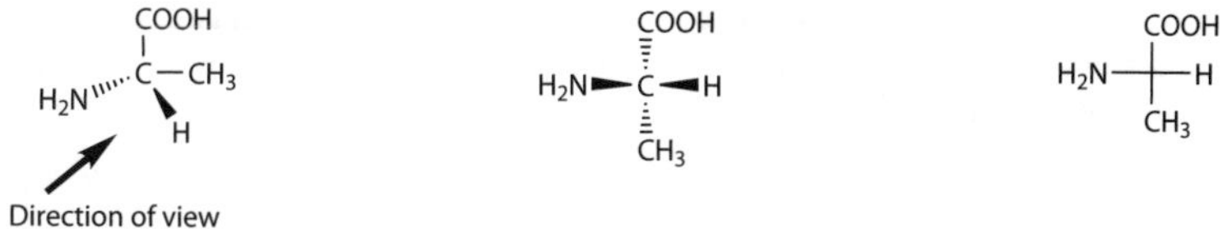

Direction of view

Bonds shown as solid wedges project towards the viewer; those shown as dashed wedges project away from the viewer

Bonds shown as solid wedges project towards the viewer; those shown as dashed wedges project away from the viewer

Fischer projection of the molecule

One of the enantiomers of the amino acid, alanine. The central carbon atom is a chiral carbon, with four different substituents attached (CH_3, H, COOH and NH_2)

Figure 4.2 Two-dimensional representations of enantiomers.

- The direction and the extent to which an enantiomer rotates the plane of plane-polarised light can be measured using a *polarimeter*.
- Enantiomers that rotate the plane of plane-polarised light to the left (that is, anticlockwise) are designated as the (−) forms, and are referred to as *levorotatory*; those that rotate the plane to the right (that is, clockwise) are designated as the (+) forms, and are referred to as *dextrorotatory*. The (+) enantiomers were once labelled as the *d*-forms and (−) enantiomers as the *l*-forms.
- A mixture that contains equal amounts of both the (+) and (−) forms of a compound is referred to as a *racemic mixture* or alternatively as a *racemate*.
- A racemic mixture of enantiomers is indicated by the prefix (±)-, and on the older system of nomenclature, using the prefix *dl*-.
- The drug, L-DOPA, used in the treatment of Parkinson's disease, has the enantiomer D-DOPA (Figure 4.3), which is biologically inactive. L-DOPA and D-DOPA are mirror images of one another and cannot be superimposed.
- The sets of stereoisomers that exist in molecules that possess two or more chiral carbons will include pairs of stereoisomers that are non-superimposable and are not enantiomers.
- The drug ephedrine contains two chiral carbons and so has four possible stereoisomers (Figure 4.4). These four stereoisomers

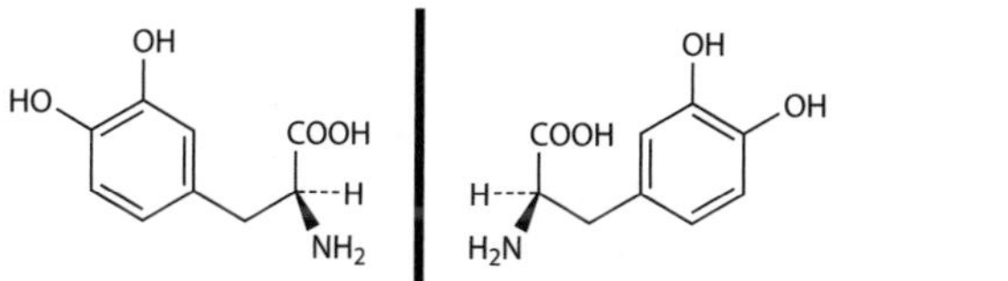

L-DOPA

D-DOPA

Figure 4.3
Stereoisomers.

The structure shown to the right is a mirror image of the one on the left; the mirror plane lies as an imaginary wall separating the two structures, perpendicular to the plane of the page (here shown as a vertical bar)

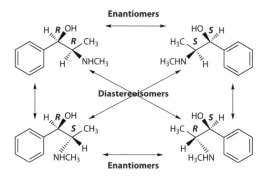

Figure 4.4
Stereoisomers of ephedrine (lower images) and pseudoephedrine (upper images).

represent two pairs of enantiomers (Figure 4.4: upper left and right images, and lower left and right images). The stereoisomers, shown top left and lower left, are not superimposable and they are not enantiomers; they are *diastereomers* (also sometimes known as *diastereoisomers*). The other three pairs of stereoisomers (top left and lower right, top right and lower left, and top right and lower right) are also diastereomers.

- Molecules that contain two or more chiral centres and that also possess an *internal mirror plane* give rise to a reduced number of stereoisomers. The existence of the internal mirror plane means that two of the possible stereoisomers will in fact be identical because the two halves of the molecule are symmetrical and superimposable. This non-optically active member of the set of stereoisomers is known as the *meso-form*. Tartaric acid, for example, has two chiral centres, but exists only as three stereoisomers (Figure 4.5).

Fischer's convention of stereochemical nomenclature

The 19th-century Nobel prize-winning chemist Emil Fischer proposed a way to designate the stereochemical configuration of the chiral centres in sugars using the structure of glyceraldehyde as a reference.

With the carbon chain of glyceraldehyde oriented vertically, with the lowest numbered carbon – the aldehyde carbon – at the

A

B

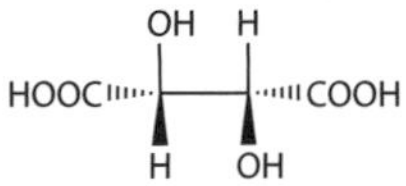

A and B are stereoisomers that are non-superimposable mirror images of one another and so are enantiomers

C

D

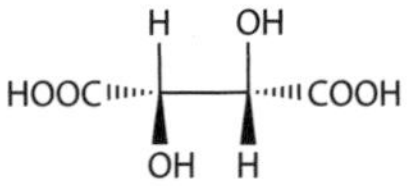

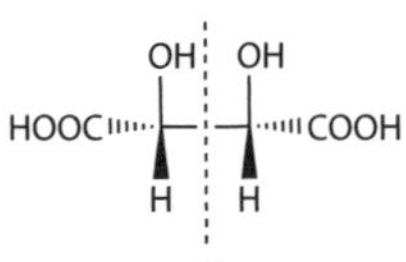

C and D show drawings of two mirror-image structures that are superimposable. The dashed line shows the internal plane of symmetry that bisects the molecule, allowing opposite ends to be superimposed. C and D thus show the same isomer of tartaric acid, which is referred to as meso-tartaric acid

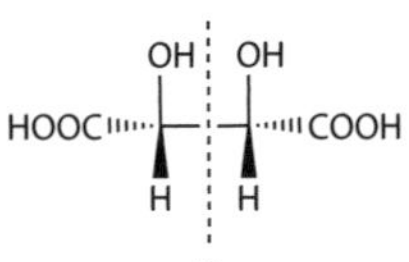

Figure 4.5 Stereochemistry of tartaric acid.

top, the substituents on the chiral carbon are drawn so that the groups projecting away from the viewer (into the page) are drawn as connected via the vertical bonds, and the groups that project towards the viewer (out of the page) are drawn as connected via the horizontal bonds (Figure 4.6). Inspection of the orientation of the substitutents about the chiral carbon then allows the stereochemistry to be specified: if the main (horizontally bonded) substituent (here, the hydroxyl) is drawn to the left of the chiral carbon, the enantiomer shown is designated the L-enantiomer, and if this group is drawn to the right of the chiral carbon, the enantiomer shown is designated the D-enantiomer. (NB: we use small capital letters here and not large ones.)

For most types of organic compound, the *Fischer system* of stereochemical nomenclature has been replaced by the one devised

D-glyceraldehyde drawn with wedge and dashed bonds

D-glyceraldehyde drawn according to Fischer's convention

Fischer projection of D-glyceraldehyde

L-glyceraldehyde drawn with wedge and dashed bonds

L-glyceraldehyde drawn according to Fischer's convention

Fischer projection of L-glyceraldehyde

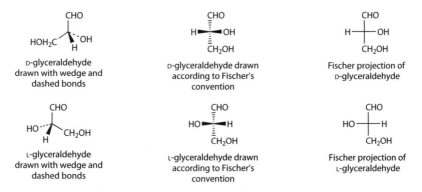

Figure 4.6 Fischer's system for representation of chiral structures and chiral nomenclature.

Figure 4.7
Fischer nomenclature for amino acids.

L-alanine drawn with wedge and dashed bonds

Fischer projection of L-alanine

Tip

Don't mix up *l-* and L- or *d-* and D-; the lower case letters refer to the optical activity of a compound, while the small capital letters designate particular enantiomeric forms on the Fischer system of chiral nomenclature.

by Cahn, Ingold, and Prelog. The Fischer system, however, is still retained to describe the stereochemistry of amino acids and sugars (Figure 4.7).

The *Cahn–Ingold–Prelog system* of stereochemical nomenclature has chiral centres designated *R* or *S*. The rules followed to determine chirality are detailed below:

1. Assign a priority to each atom attached to the chiral centre:
   a. First assign priority on the basis of atomic number, giving highest priority (priority 1) to the atom of highest atomic number and lowest priority (priority 4) to the atom of lowest atomic number.
   b. If any of the atoms attached to the chiral carbon have the same atomic number, prioritise on the basis of the atomic numbers of the atoms that are one bond removed from the chiral carbon (and repeat until all four groups have been assigned a priority order).
   c. In the case of atoms with multiple bonds (C=O, for example), these are treated as having the equivalent number of single-bonded species (C=O, therefore, would be treated as O-C-O).
2. View the molecule from the side opposite the lowest-priority atom/substituent (priority 4) with this group drawn projecting backwards away from you, leaving the other three atoms/groups (priorities 1–3) drawn in the foreground as the three spokes of a steering wheel.
3. If the three foreground groups have a clockwise priority order, the enantiomer shown is the *R*-enantiomer (the *R* being an abbreviation for *rectus,* the Latin for right); if they have an anticlockwise priority order, the enantiomer shown is the *S*-enantiomer (*S* being an abbreviation of *sinister,* the Latin for left).

Figure 4.8 shows the structure of (*R*)-alanine, and the more complex case of salbutamol is shown in Figure 4.9.

(R)-alanine (at left). The lowest-priority group, H, is put to the back, and the remaining three groups follow the clockwise priority order: NH$_2$, COOH, CH$_3$ (shown at right). The NH$_2$ group has highest priority (priority 1) because the atomic number of N > atomic number of C. COOH has a higher priority than CH$_3$ because the attached Os have higher atomic number than the Hs in the methyl group.
NB: Following Fischer's convention, (R)-alanine would be reported as D-alanine.

Figure 4.8 Cahn–Ingold–Prelog system of stereochemical nomenclature – alanine.

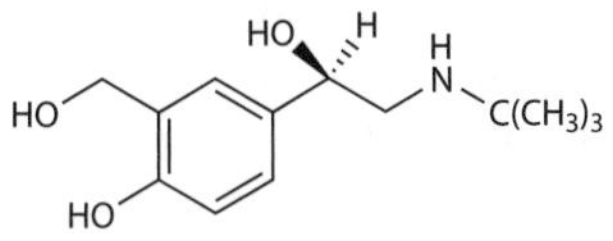

Figure 4.9
Cahn–Ingold–Prelog system of stereochemical nomenclature – salbutamol.

(R)-salbutamol. The lowest-priority group, H, is put to the back, and the remaining three groups follow the clockwise priority order: OH, CH$_2$N, aromatic C.

Conformational isomers

Conformational isomers (often also referred to just as *conformers*) are a type of stereoisomer; they are different spatial arrangements of the atoms in a molecule that can be interconverted simply by rotation about single bonds.

Where the rotation about a single bond in a molecule is restricted/hindered, because there is an energy barrier (associated with *bond strain*) to be overcome to convert one conformer to another, the energetically preferred conformers are referred to as *rotamers*.

The spatial arrangement of atoms in a given conformer can be conveniently displayed (on paper) using a *sawhorse projection*, in which the atoms or groups attached to the atoms involved in a single bond are considered, with the groups that are attached to the front and back carbons drawn with their connecting bonds set at 120° to one another (Figure 4.10). An alternative representation shows the molecule in *Newman projection*, viewed along a given single bond and with the attached groups drawn as the spokes of two superimposed 'steering wheels'.

KeyPoints

- A racemic mixture (or racemate) is one that contains equal amounts of two enantiomers.
- Racemic mixtures may be variously indicated by means of the prefixes *dl*-, DL- or (±).

KeyPoints

- Conformational isomers (or conformers) are stereoisomers that have different spatial arrangements of their constituent atoms that are interconverted by rotation about single bonds.
- In open-chain molecules, the most energetically favourable conformers are those with staggered bonds and the most unfavourable conformers are those with eclipsed bonds.
- In non-aromatic cyclic molecules, the most energetically favourable conformers are the chair forms and the most unfavourable conformers are the boat forms.

In the stimulant and appetite suppressant, ephedrine, the preferred conformer will have the bulky methylamine ($NHCH_3$) and phenyl (Ph) groups oriented so that they point in diametrically opposite directions (as shown in Figure 4.10C and D). With the groups arranged in this way, the methylamine and phenyl groups are said to adopt an *anti* arrangement. In the same conformer, the OH and $NHCH_3$ groups are oriented at 60° to one another, and are said to adopt a *gauche* arrangement. The most energetically unfavourable conformer in an open-chain molecule will be one in which the substituents on one bonded carbon are oriented in the same sense as their counterparts on the other bonded carbon (as shown for ephedrine in Figure 4.10E and F). In such conformers the (overlapping) bonds are said to be *eclipsed*.

In eclipsed conformers the atoms/ subsitutents attached to bonded carbons will approach closer than the sum of their van der Waals' radii, and this leads to an interatomic repulsion, giving rise to *steric hindrance*.

Stereoisomers of cyclic molecules

- The stereoisomers of cyclic molecules can be depicted using wedge and dashed bonds (Figure 4.11A), or else as *Haworth projections* (Figure 4.11B).
- The energetically preferred conformers for (non-aromatic) cyclic molecules have all bonds associated with the ring atoms in a staggered arrangement (as far apart as they can possibly be). This arrangement is achieved in the *chair* conformer (Figure 4.11C).
- If the bonds are arranged so that they are eclipsed (and are then as close as they can possibly be), a *boat* conformer results (Figure 4.11D).
- The conformer midway between the chair and boat forms is known as the *half-chair*. In this form, there are five of the six ring atoms in a plane, and this causes distortion of the bond angles (so that they deviate, unfavourably, from the tetrahedral angle, 109°).
- The *twist boat* conformer is, as the name suggests, a twisted form of the boat conformer. It is slightly more stable than the half-chair conformer because the angle strain is relieved by the twist in the structure.
- In cyclic molecules, it is energetically more favourable to have the bulkier ring substituents projecting with an *equatorial* orientation (the bonds to these groups being coplanar with the

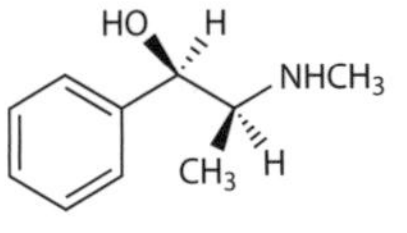

(A) Molecular structure of ephedrine

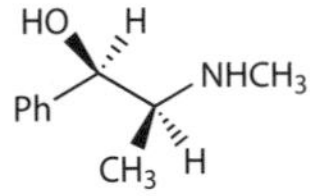

(B) Simplified representation of the structure

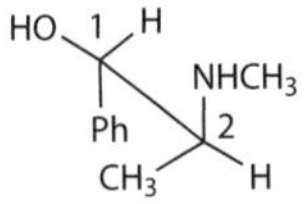

(C) Sawhorse projection (with the molecule viewed along the bond from C2 to C1). The bonds to C1 and those to C2 are all staggered

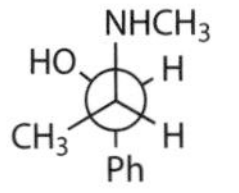

(D) Newman projection (with the molecule viewed along the C2–C1 bond). The NHCH₃ and phenyl groups are here shown in a staggered arrangement

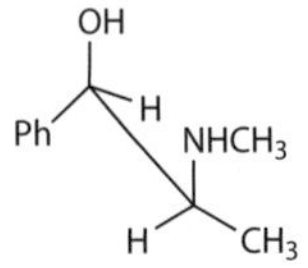

(E) Sawhorse projection (with the molecule viewed along the C2–C1 bond). The bonds to C1 and those to C2 are eclipsed

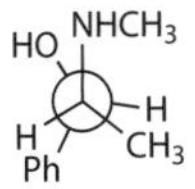

(F) Newman projection (with the molecule viewed along the C2–C1 bond). The bonds to C1 and C2 are now eclipsed, giving rise to unfavourable close contacts between the substituents

Figure 4.10 Two-dimensional representations of ephedrine.

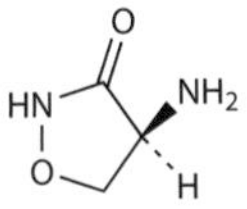

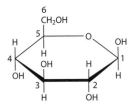

(A) The antitubercular drug, cycloserine, with stereochemistry at the chiral carbon shown by means of wedge and dashed bonds

(B) Haworth projection of β-glucose molecule, clearly showing that the substituents on the 6-membered ring do not lie in the ring plane

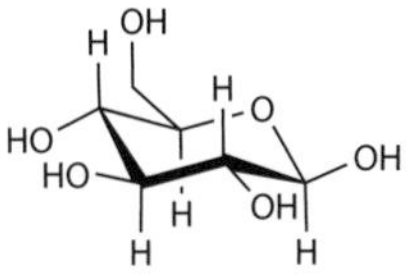

(C) The ⁴C₁ chair form of β-glucose. In this form, atom C4 lies above and atom C1 below the plane defined by C2, C3, C5 and O

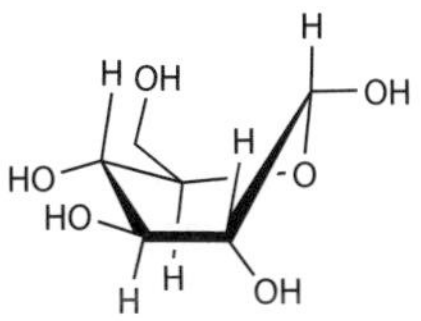

(D) The boat form of α-glucose. In this form, atoms C2, C3, C5 and O lie in a plane, and atoms C1 and C4 are above the plane, on the same side of the ring

Figure 4.11 Two-dimensional representations of stereochemistry in cyclic molecules.

ring), and to have the smaller substituents – for which steric hindrance is less of an issue – with *axial* orientations (the bonds to these groups oriented perpendicular to the ring plane).

Geometric isomers

Geometrical isomers arise when groups of atoms are arranged asymmetrically about a *double bond*. There are two possible configurations about any given double bond, one of which is identified as the *E*-isomer (the *E* here being an abbreviation of *entgegen*, the German for opposite) and the other the *Z*-isomer (the *Z* here being an abbreviation for *zusammen*, the German for together).

The *E*- and *Z*-isomers are distinguished according to whether the two higher-priority groups lie, respectively, on the opposite sides of the double bond, or on the same side (Figure 4.12).

The priority of groups is determined in exactly the same way as for determining the absolute stereochemistry of chiral centres: higher priority is given to atoms with higher atomic number, and if the atoms bonded directly to the double bond have the same atomic number, the priority is determined by considering the atomic numbers of the atoms at one bond distant from the double bond, and so on, until the priority is settled.

An older (much less useful) system of nomenclature employed in naming geometrical isomers used the labels *cis*- and *trans*- to distinguish the isomers. With this system of nomenclature, the two carbons involved in the double bond were required to have one substituent in common. The *cis*-isomer has the common substituent located on the same side of the double bond; the *trans*-isomer has the common substituent located on opposite sides of the double bond (Figure 4.12).

KeyPoints

- (*R*) and (*S*) prefixes specify the arrangement of groups attached at chiral carbons.
- (*E*) and (*Z*) prefixes specify the arrangement of groups attached to the carbons involved in double bonds.

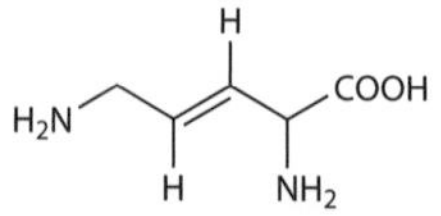

E-dehydroornithine. For the groups attached to the leftmost alkene carbon, priority 1 is given to the CH_2 group, and at the other alkene carbon, priority 1 is assigned to the $CH(NH_2)COOH$ group. The two priority 1 groups lie on opposite sides of the double bond and so it is the *E*-isomer that is drawn here. On the older system of nomenclature this isomer would also be referred to as *trans*-dehydroornithine, since the substituent common to both alkene carbons (H) is arranged on opposite sides of the double bond.

Figure 4.12 Geometric isomers – dehydroornithine.

The *E*- and *Z*-isomers of an alkene are not readily interconverted, since this requires that the π-bond must be broken so that the groups at either end of the remaining σ-bond can rotate, and the π-bond can then be reformed – this whole process being energetically highly unfavourable.

Topological isomers

Complex macromolecules like nucleic acids, proteins and polysaccharides exist as complex three-dimensional structures with a specific *topology*. In such cases, the different isomers are known as *topological isomers* or *topoisomers*. The polynucleotide chains in DNA, for example, adopt a right-handed helical conformation (wound together to form the famous double helix: Figure 4.13A) while the polypeptide chain in each subunit of haemoglobin is locally folded so as to form a set of (eight) right-handed *alpha-helices* (Figure 4.13B). Note, therefore, that these biomolecules are polymers involving chiral molecules (amino acids in proteins, sugars in DNA) and that their macromolecular forms are also chiral, forming helices that exhibit the same handedness as seen in spiral staircases.

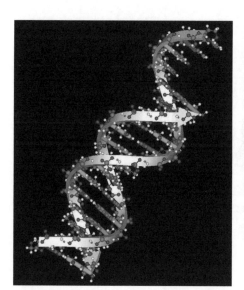

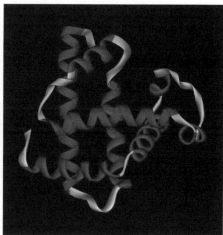

Cartoon representation of one subunit of haemoglobin showing the alpha-helices (highlighted)

B

Molecular model of DNA and a superimposed cartoon ribbon highlighting its nature as a double helix

A

Figure 4.13 Topoisomerism in DNA and haemoglobin.

Tips

- Even very subtle changes in a molecule's stereochemistry can cause marked changes in its properties. While (–)-(S)-limonene (below left) is responsible for the distinctive taste of lemons, (+)-(R)-limonene (below right) is responsible for the taste of oranges!

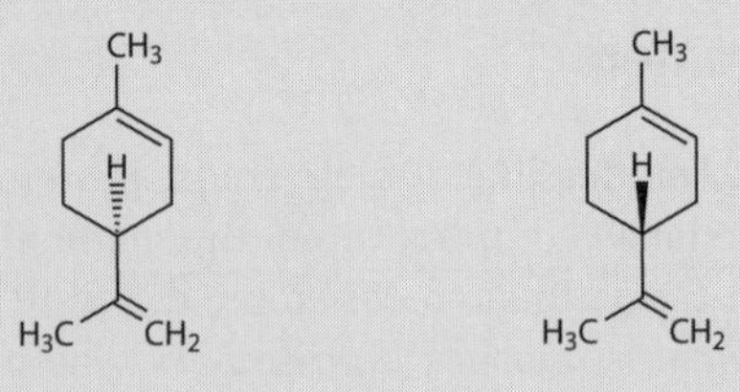

- The enantiomers of chiral drug molecules very often have quite different pharmacological activities. In the antidepressant drug, citalopram, for example, the (S)-enantiomer (below left) lacks the adverse side effects of the (R)-enantiomer (below right).

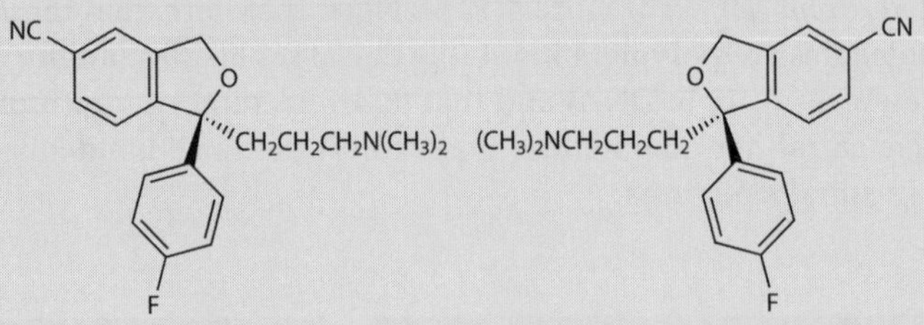

Self-assessment

1. The molecule shown below:

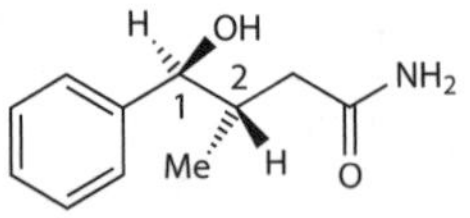

has the absolute configuration:
   a.   1S,2R
   b.   1R,2S
   c.   1S,2S
   d.   1R,2R

2. In the structure below:

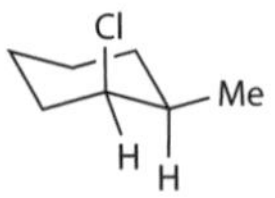

the methyl group and the chlorine atom are:
   a.   *cis* to one another with the methyl equatorial and the chlorine axial
   b.   *trans* to one another with the methyl axial and the chlorine equatorial

c. *cis* to one another with the methyl axial and the chlorine equatorial
d. *trans* to one another with the methyl equatorial and the chlorine axial

3. **A molecule with four different stereocentres can exist in how many different stereogenic forms?**
a. 4
b. 8
c. 12
d. 16

4. **Stereoisomers differ from each other in their:**
a. composition
b. constitution
c. configuration
d. steric hindrance

5. **The molecule shown below:**

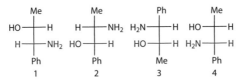

has the absolute stereochemistry:
a. 1*E*,2*Z*
b. 1*Z*,2*E*
c. 1*E*,2*E*
d. 1*Z*,2*Z*

6. **Which of the following statements is true?**
a. Dextrorotatory enantiomers always have the *R* absolute configuration.
b. Laevorotatory enantiomers always have the *R* absolute configuration.
c. Enantiomers may be optically inactive due to internal compensation and exist as meso forms.
d. The direction of rotation of the plane of plane-polarised light gives no indication of the absolute configuration of a molecule.

7. **Which of the following Fischer projections represent a pair of enantiomers?:**

	Me		Me		Ph		Me
HO—	—H	H—	—NH$_2$	H$_2$N—	—H	HO—	—H
H—	—NH$_2$	HO—	—H	HO—	—H	H$_2$N—	—H
	Ph		Ph		Me		Ph
	1		2		3		4

a. 1 and 2
b. 3 and 4
c. 1 and 4
d. 2 and 3

8. The absolute configuration of the molecule below:

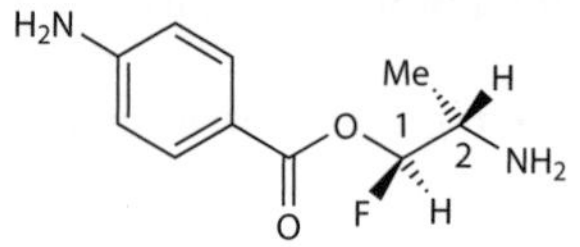

    is given by:
    a. 1*S*,2*R*
    b. 1*R*,2*S*
    c. 1*S*,2*S*
    d. 1*R*,2*R*

9. Which of the following sugars has the configuration (2R,3S,4R)?

    a.

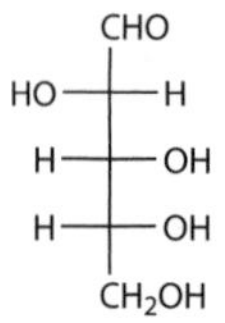

    b.

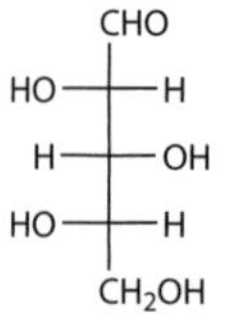

    c.

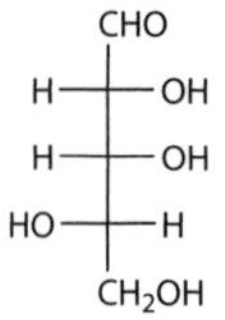

    d.

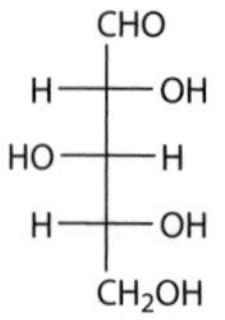

10. Which of the following compounds represent a pair of enantiomers?

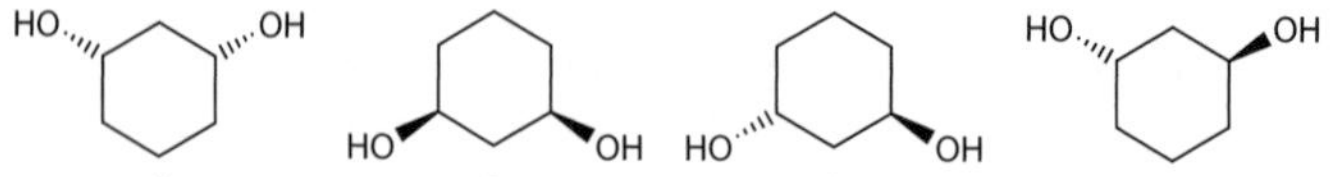

    a. 1 and 2
    b. 2 and 3
    c. 3 and 4
    d. 2 and 4

chapter 5
Chemical reaction mechanisms

Overview

After learning the material presented in this chapter you should:
- appreciate what a reaction mechanism shows
- understand the difference between a nucleophile and an electrophile
- be able to determine whether a molecule or group is nucleophilic or electrophilic
- be able to use the curly arrow notation to draw a simple reaction
- be able to use the 'fish hook' notation to draw a simple radical reaction
- consider the limitations of a reaction mechanism.

- A reaction mechanism is a step-by-step account of what happens during a chemical reaction.
- The reaction mechanism shows all of the chemical species that are formed during the reaction, many of which will not be observable experimentally.
- Reaction mechanisms show which bonds are broken and which are formed during the course of the reaction using the curly arrow notation.
- Intermediates are chemical species formed during a reaction which are not products.

Nucleophiles and electrophiles

- Molecules that are electron-rich with either a free pair of electrons or a π bond that they can donate are termed nucleophilic (nucleus loving). Since nucleophiles donate a pair of electrons they are also Lewis bases.
- Molecules that are electron-deficient and are able to accept a pair of electrons from a nucleophile are termed electrophilic (electron loving). Since electrophiles accept a pair of electrons they are also Lewis acids.

KeyPoints

- All Lewis bases are nucleophiles, but a strong base may not be a good nucleophile and vice versa.
- All Lewis acids are electrophiles, but a strong acid may not be a good electrophile and vice versa.

Reaction mechanisms

- To write a reaction mechanism you first need to identify the nucleophile and the electrophile.
- Each step in a reaction will have a nucleophilic component and an electrophilic component; this will be where a pair of electrons is starting from (the nucleophile) and where it will be going (the electrophile).
- The non-bonding, lone pairs of electrons on an atom can be used to form a new bond. Conversely, non-bonding, lone pairs of electrons can be created through the breaking of a pre-existing bond. When drawing curly arrows to show the progression of a reaction, these arrows can therefore begin or terminate on individual atoms.
- When drawing a reaction mechanism it is important to be both accurate and precise in where the curly arrow is coming from and where it is pointing (Figure 5.1).

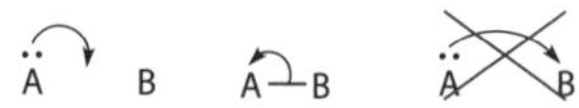

Figure 5.1
Positioning of curly arrows.

Tips

- Never forget the octet rule: count your electrons!
- Carbon (or any other second-row element) will never form five bonds. It cannot accommodate 10 electrons.
- Any mechanism that creates a five-bonded carbon is wrong.
- If forming a new bond would result in too many bonds being present on an atom then there must be an accompanying curly arrow to break one of the original bonds at that atom.

- If a bond is being formed the arrow will point between the two atoms that will form the bond. If a bond is being broken the arrow will point to one of the atoms that formed the bond originally. A bond will not be formed by electrons moving from one atom to another.
- When the nucleophile and the electrophile are in the same molecule, this is termed an intramolecular process. When the nucleophile and electrophile are on different molecules this is an intermolecular process.
- Intramolecular reactions are usually faster than intermolecular reactions since the two reacting centres are being held in close proximity by the rest of the molecule.

Curly arrows in reaction mechanisms

- A lone pair is used to form the bond. The arrow starts on the oxygen and ends between the two atoms that will form the bond (Figure 5.2).

Figure 5.2
Single-arrow movement of electrons.

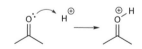

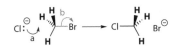

Figure 5.3
Two-arrow movement of electrons (resonance).

Figure 5.4
Two-arrow movement of electrons (reaction).

Figure 5.5
Pairs of electrons in reaction mechanisms.

- Oxygen has formally *lost* an electron, so it becomes *positively* charged (Figure 5.2).
- A lone pair moves to form a new bond (arrow a in Figure 5.3). The curly arrow ends in the middle of the bond to be formed.
- If a double bond is formed, then the central carbon atom has too many electrons, so another bond needs to break (arrow b). The two electrons of the light blue bond retire as a lone pair to the oxygen.
- The oxygen has formally *gained* an electron, making eight electrons in total, and becomes *negatively* charged.
- A lone pair from chlorine attacks the electrophile (arrow a in Figure 5.4), forming the blue bond between the chlorine and the central carbon. The curly arrow starts at chlorine and ends between the chlorine and the carbon.
- The carbon now has to displace another bond. Otherwise it would have more than eight electrons. The light blue bond breaks, and its two electrons retire on bromine as a lone pair (arrow b in Figure 5.4).
- The curly arrow can start from either the pair of electrons on the chlorine (Figure 5.5) or from the negative charge.
- As structures get more complicated, you will not want to draw out all the pairs of non-bonding electrons on all the atoms.
- In a mechanism, all charges on both sides of the equation must balance; electroneutrality must be maintained. If there is no charge separation in the starting reagents but the reaction leads to the formation of a negative charge in one of the products, then a complementary positive charge must also have been created and vice versa.

Radicals in reaction mechanisms

- In radical reactions it is not the movement of pairs of electrons that we are interested in but the movement of single, unpaired electrons.
- For reaction mechanisms involving radicals the single-headed arrow or 'fish hook' is used (Figure 5.6).

Figure 5.6
Different types of curly arrows.

Movement of a   Movement of a
pair of electrons   single electron

Fish hooks in reaction mechanisms

- The unpaired electron on bromine attacks the electrophilic alkene (arrow a in Figure 5.7); this causes fission of the π bond.
- One electron from the π bond moves towards the bromine (arrow b), forming the blue bond. The other π electron moves to the end of the double bond (arrow c) and localises on the carbon.
- In effect the radical has attacked the double bond and moved through the π system.
- As a shorthand, arrow b will sometimes be omitted from mechanisms, its presence being implied by the other two arrows.

Figure 5.7
Movement of an electron using fish hooks.

Chemical reactivity

- When a molecule contains a number of functional groups it may be possible to write more than one mechanism. Which reaction occurs will depend on the relative rates of reaction for each mechanism.
- Some functional groups will be more stable under certain reaction conditions and will require more forcing conditions to react. It is important to consider the reaction conditions when drawing a mechanism. Solvent, reaction temperature and pH all affect chemical reactivity.
- A reaction mechanism does not take into account the effect of substituents adjacent to the reaction centre which may influence the course of a reaction. Steric hindrance can prevent a reaction from occurring even though a mechanism with appropriate curly arrows can be drawn. Hydrogen bonding can allow a molecule to adopt a specific conformation that might allow or prevent a reaction from occurring.
- A reaction mechanism will not tell you whether a reaction will work or not. The only way to know this for certain is to carry out the reaction experimentally.
- When drawing a mechanism it is essential to look at the molecule as a whole. Where there are a number of sites which could react, each should be evaluated independently to determine the product of the reaction.

Tip

It is not necessary to memorise lots of different reaction mechanisms. If you understand the nature of the reagents and the various functional groups and can push curly arrows, you can complete any reaction mechanism, no matter what the molecule might look like.

Self-assessment

1. **In reaction mechanisms, curly arrows are used to show?**
a. the movement of a single electron
b. the movement of a pair of electrons
c. the movement of atoms within the molecule
d. the direction of attack of the approaching species

2. **What is missing from the following reaction mechanism?**

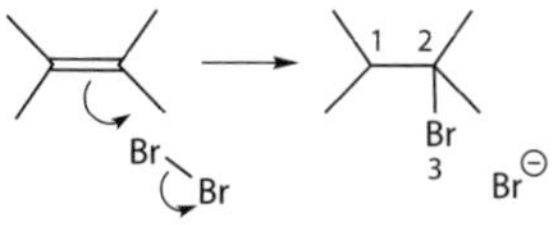

a. a positive charge on carbon 1
b. a positive charge on carbon 2
c. a double bond between carbon 1 and carbon 2
d. a positive charge on bromine 3

3. **In the reaction given below, what is missing from the sulfur atom in the product?**

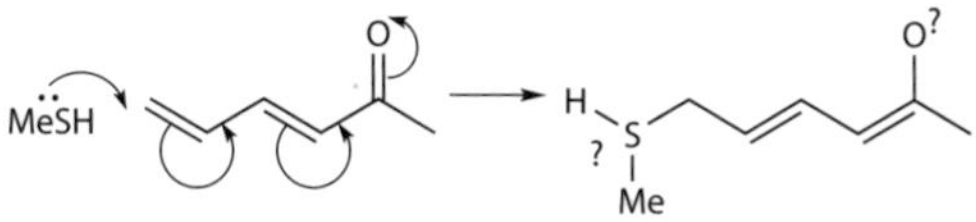

a. an additional non-bonding pair of electrons
b. a formal negative charge
c. a formal positive charge
d. nothing is missing since sulfur has eight valence electrons

4. **Which of the following species is not a nucleophile?**
a. sodium ion
b. chloride ion
c. water molecule
d. cyanide ion

5. **What is missing from the following reaction mechanism?**

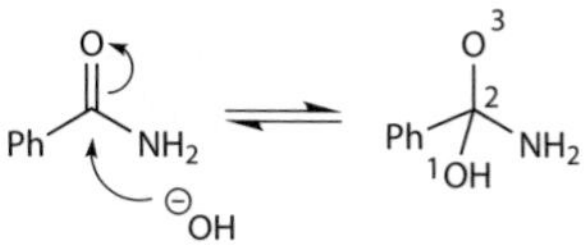

a. a positive charge on oxygen 1
b. a negative charge on oxygen 1
c. a negative charge on carbon 2
d. a negative charge on oxygen 3

6. Which of the following species is not an electrophile?
a. NO_2^+
b. Br^+
c. BF_3
d. $CH_2=CH_2$

7. Which of the following mechanistic steps correctly shows the positioning of curly arrows in a simple substitution reaction?

a.

b.

c.

d.

8. Which of the following arrow types denotes a reversible reaction?
a.
b.
c.
d.

9. Which of the following statements about the drawing of reaction mechanisms is not true?
a. The charges on both sides of the reaction arrow must balance.
b. Electrons will always flow in the direction of the curly arrow.
c. Each step in a reaction will have a nucleophilic and an electrophilic component.
d. A reaction will be reversible if you can draw the curly arrows for the reverse process.

10. **Which of the following mechanistic steps correctly shows the positioning of curly arrows in a simple elimination reaction?**

a.

b.

c.

d.

chapter 6
Chemistry of electrophiles and nucleophiles

Overview

After learning the material presented in this chapter you should:

- understand electrophilic addition, nucleophilic substitution and elimination reactions
- understand and be able to explain the mechanisms for these reactions
- understand the stereochemistry and regioselectivity in these reactions
- be able to consider the stability and reactivity of the substrates and intermediates involved in these reactions
- be able to discuss carbocation stability
- understand and be able to explain how the nucleophile, base, solvent, leaving group and substrate affect the course of a reaction and the mode of a reaction.

Addition and substitution reactions

- An addition reaction is one in which two reactants combine to make a single product. Every atom from the two reactants is present in the product.

$$A + B \longrightarrow AB$$

- A substitution reaction is one in which one reactant displaces an atom or group from another reactant. This exchange results in the formation of two new products.

$$A + BC \longrightarrow AB + C$$

- Electrophiles and nucleophiles undergo both addition and substitution reactions (Figure 6.1). The specific reaction that an electrophile or nucleophile will undergo will depend on the nature of the electrophile/nucleophile, the nature of the molecule it is reacting with and on the reaction conditions.

Electrophilic addition

- The π bond of an alkene is electron-rich with high electron density located above and below the plane of the carbon framework.

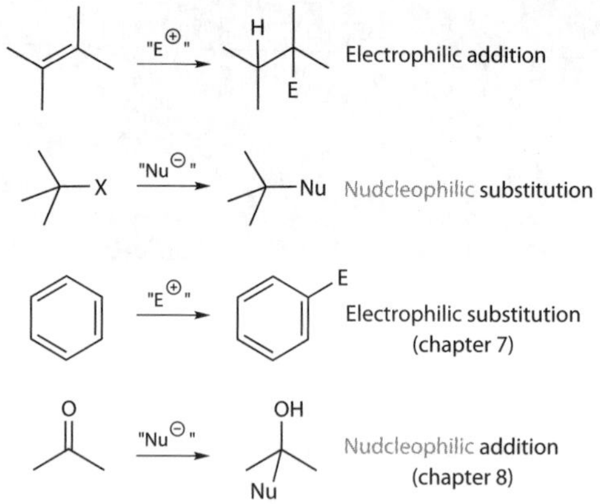

Figure 6.1

Types of addition and substitution reactions.

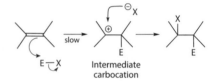

Electrophilic addition

Nudcleophilic substitution

Electrophilic substitution
(chapter 7)

Nudcleophilic addition
(chapter 8)

Figure 6.2

Generalised mechanism for electrophilic addition.

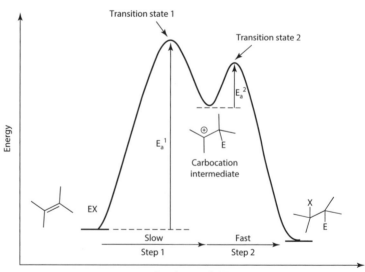

Intermediate
carbocation

Figure 6.3
Reaction profile
for electrophilic
addition.

Transition state 1

Transition state 2

E_a^2

Energy

E_a^1

Carbocation
intermediate

EX

Slow
Step 1

Fast
Step 2

Reaction coordinate

- The two electrons of the π bond are only weakly held by the two carbon atoms, making alkenes nucleophilic in nature.
- When an alkene reacts with an electrophile it is the electrons from the π bond that attack, forming a new bond between the electrophile and one of the carbon atoms of the double bond (Figure 6.2).

- The electrons from the σ bond remain unchanged and covalently bond the two carbon atoms together.
- The process of breaking the double bond results in a positive charge developing on the carbon not attached to the electrophile; an intermediate carbocation is formed.
- The counter ion partner of the electrophile then reacts rapidly with the electron-deficient carbocation to give the addition product (Figure 6.3).

Changes in hybridisation during electrophilic addition

- Initially, both carbon atoms of the alkene are sp² hybridised.
- To react with the electrophile the hybridisation of one of the atoms must change to sp³ to allow a new σ bond to be formed. The new hybridisation mode allows the four pairs of electrons in the bonds surrounding the carbon to minimise their mutual repulsion.
- The carbon at the other end of what was the double bond maintains sp² hybridisation with an empty p orbital. The nucleophilic X⁻ is able to attack the planar sp² hybridised carbocation from either above or below the plane of the carbon framework.
- To form a bond to the nucleophile the hybridisation must again change to one of sp³ hybridisation.
- During the course of electrophilic addition both carbon atoms of the alkene change from sp² to sp³ hybridisation (Figure 6.4).

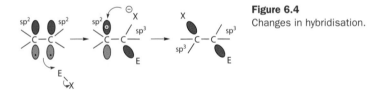

Figure 6.4
Changes in hybridisation.

Stereochemistry of electrophilic addition

- One implication of having a planar sp² hybridised carbocation intermediate is that when the carbocation bears three different substituents, addition of the nucleophile to either the top face (arrow a in Figure 6.5) or the bottom face (arrow b) of the carbocation will yield different products.
- The two molecules formed are mirror images of each other. They are enantiomers and, since the nucleophile has an equal probability of adding to the top face as it does to the bottom face, the two enantiomers will be formed in equal amounts, and a racemic mixture will result.

Figure 6.5
Nucleophilic attack at the sp²
hybridised carbon.

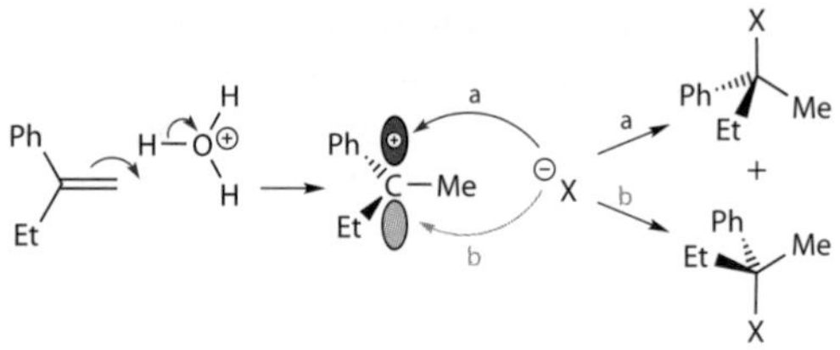

- The two faces of the carbocation are termed enantiotopic. The addition of a nucleophile to each of the two faces leads to the formation of different enantiomers.

Examples of electrophilic addition

- The mechanism for all electrophilic addition reactions begins with the alkene attacking the electrophile, generating a carbocation intermediate which is then quenched to give a fully saturated product.

Addition of hydrogen halides

- Although different mechanisms can be drawn for a reaction, the actual mechanism that a reaction proceeds by can only be confirmed experimentally.
- For the electrophilic addition of HCl to an alkene (Figure 6.6):
  - There is no reaction between an alkene and aqueous sodium chloride. This implies the chloride ion (Cl⁻), sodium ion (Na⁺) and/or water are not able to initiate the reaction: they cannot be reacting first.
  - The addition of acid to the previous reaction mixture initiates the reaction immediately. This indicates that H_3O^+ is reacting in the first step of the reaction and not Cl⁻.
  - Increasing the concentration of acid speeds up the reaction. No reaction is observed under basic conditions. This implies that the rate of the reaction is dependent on the concentration of acid, that is:

$$Rate = k \, [\text{alkene}] \, [H^+]$$

  H_3O^+ must be involved in the rate-determining first step of the mechanism.

Figure 6.6
Electrophilic addition of HCl to alkenes.

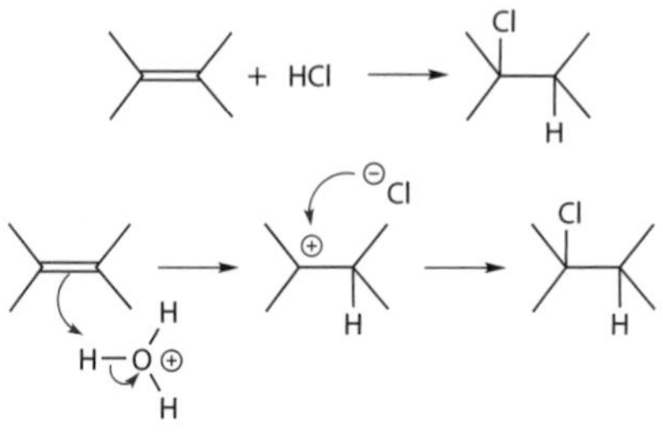

■ The addition of NaBr and NaI to the previous reaction mixture gives three products, indicating that hydrogen and chlorine are not added to the alkene in a single step. The mechanism must involve the formation of an intermediate that can react with Br⁻ and I⁻ as well as Cl⁻.

Addition of halogens

■ Bromine (Br_2) as well as chlorine (Cl_2) and iodine (I_2) react with alkenes in the same way as HCl. Delocalisation of electron density in the halogen molecule results in an electrophilic bromine and a nucleophilic bromine (Figure 6.7).

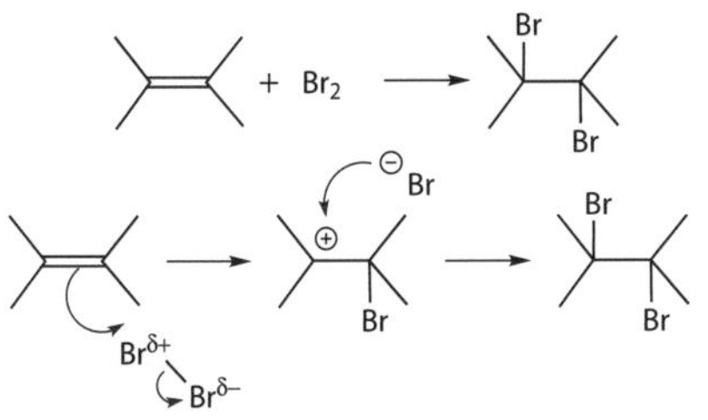

Figure 6.7
Electrophilic addition of Br_2 to alkenes.

■ When HCl is added to an alkene, the carbocation intermediate can be attacked from either the top face or the bottom face. However, the addition of bromine always results in the two bromine atoms being added opposite to each other (Figure 6.8). The two bromine atoms will never add to the same face of the alkene.

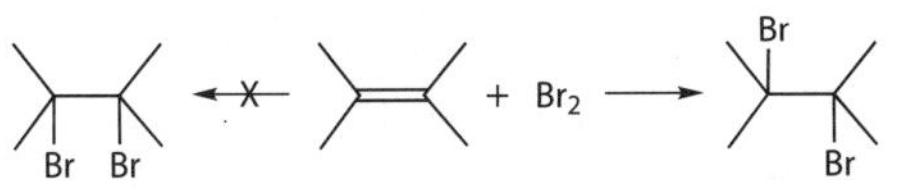

Figure 6.8
Addition of Br_2 always occurs to opposite faces of the alkene.

■ This implies that the addition of bromine does not proceed through a conventional carbocation intermediate.
■ The bromine atom has a pair of non-bonding electrons available to quench the carbocation and generate an intermediate bromonium ion (Figure 6.9).

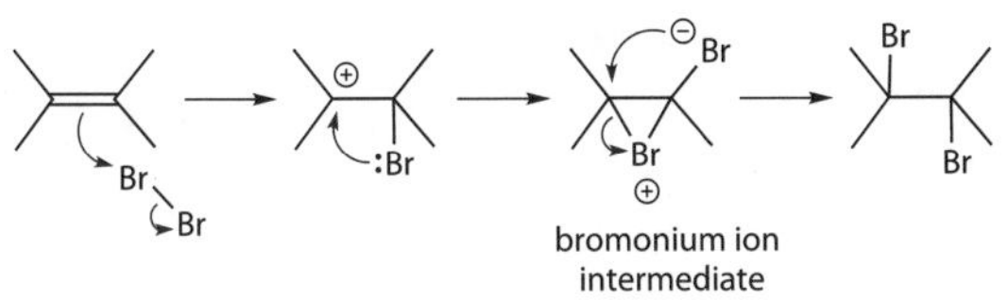

bromonium ion
intermediate

Figure 6.9
Formation of the bromonium intermediate.

KeyPoints

- Test for the presence of alkenes.
- Shaking an alkene with bromine water (orange) gives a colourless solution.
- Alkenes will decolorise bromine water.

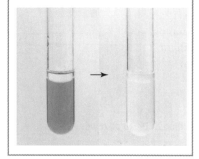

- The incoming bromide ion must then attack the bromonium ion from the opposite face to the positively charged bromine, resulting in the two bromine atoms being added *anti* to each other.
- For 1,2-disubstituted alkenes, if the carbocation formed from the initial bromine addition bears three different substituents then the final product of the reaction could be one of two possible diastereoisomers. The incoming bromide either adds opposite to the bromine previously added (arrow a in Figure 6.10) in an *anti*-addition or the bromide adds to the same side as the bromine previously added (arrow b) in a *syn*-addition.

Figure 6.10
Diastereoselectivity of Br_2 addition.

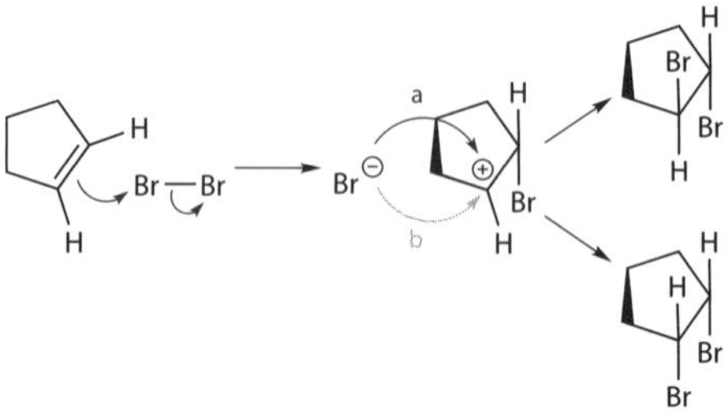

- On steric grounds you would expect the *anti*-addition product with the two bromine atoms *trans* to dominate, the incoming nucleophile preferring to attack the carbocation on the face opposite to the bulky bromine atom. However, the reaction is stereospecific, forming the *anti*-addition product exclusively; there is no *syn*-addition product formed.
- The formation of an intermediate bromonium ion is again responsible for the selectivity of this process (Figure 6.11). The opening of either end of the bromonium ion will result in the two bromine atoms being located *trans* to each other in the product.

Figure 6.11
Ring opening of the bromonium intermediate.

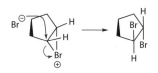

- The two faces of the carbocation are termed diastereotopic. The addition of a nucleophile to each of the faces leads to the formation of different diastereomers.

Addition of concentrated sulfuric acid

- Concentrated sulfuric acid will react with alkenes in the same way as HCl. Protonation of the alkene forms a carbocation intermediate which is then quenched by a nucleophile (Figure 6.12). In this case the nucleophile is the conjugate base of sulfuric acid and results in the formation of an alkyl hydrogen sulfate product.

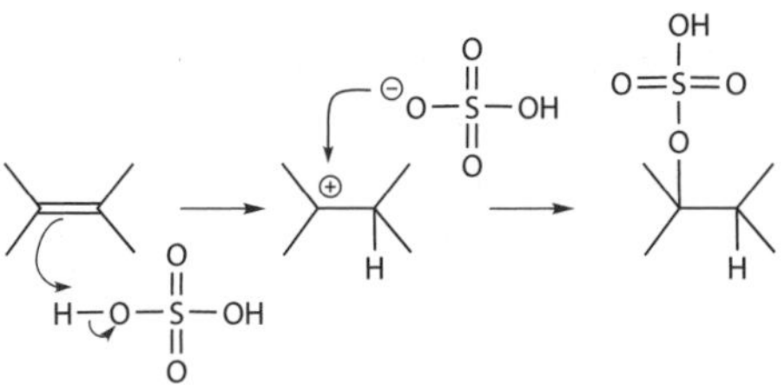

Figure 6.12
Electrophilic addition of H_2SO_4 to alkenes.

Addition of dilute sulfuric acid

- In dilute sulfuric acid the H_2SO_4 molecules are fully dissociated. The proton source for electrophilic addition is therefore H_3O^+.

$$H_2SO_4 + H_2O \rightleftharpoons HSO_4^- + H_3O^+$$

- The intermediate carbocation will now be quenched by water since the concentration of water molecules will greatly exceed the concentration of HSO_4^- ions present. The product of the electrophilic addition is an alcohol; the alkene has been hydrated (Figure 6.13).

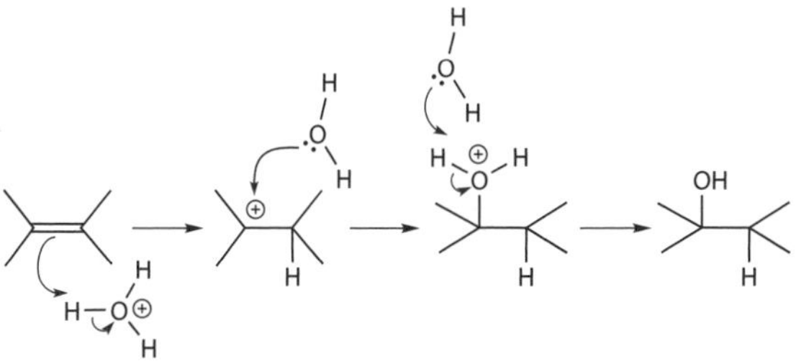

Figure 6.13
Electrophilic addition of H_2O to alkenes.

Addition of alcohols

- Under acidic conditions alcohols react with alkenes in a reaction directly analogous to the hydration mechanism (Figure 6.14). The first step of the reaction involves the protonation of the alcohol; this species is the proton source for carbocation formation in the same way that H_3O^+ is the proton source in the hydration reaction.

Figure 6.14
Electrophilic addition of MeOH
to alkenes.

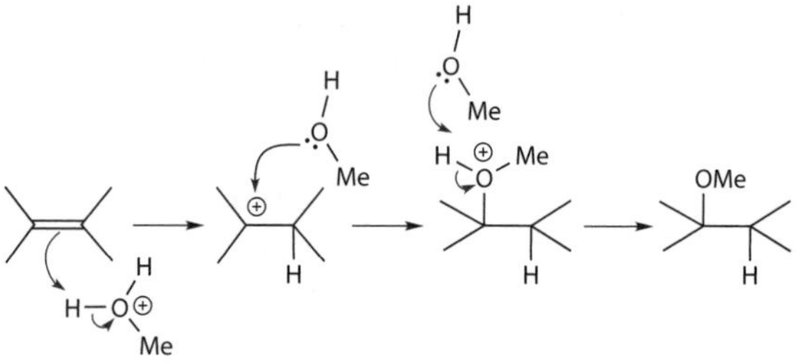

Selectivity in electrophilic addition

■ A regio*specific* reaction is a reaction where only a single structural isomer of a product is formed despite other isomers being possible. An example of a regiospecific reaction is the electrophilic addition of halogens to alkenes where only the *anti* product is observed.

■ A regio*selective* reaction is a reaction in which two or more structural isomers are formed but where one product predominates (Figures 6.15 and 6.16). An example of a regioselective reaction is the electrophilic addition of hydrogen halides to alkenes where two structural isomers are observed.

KeyPoints

Markovnikov's rule
■ When adding HX to a double bond, the hydrogen adds to the carbon which already has the most hydrogen atoms attached.
■ This is an empirical rule and will *not* always give the correct product for an addition.
■ Do not rely on Markovnikov's rule!

Figure 6.15
Electrophilic addition of HCl is regioselective.

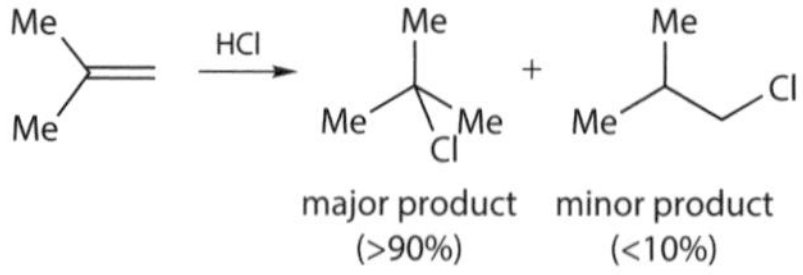

major product   minor product
(>90%)   (<10%)

Figure 6.16
Further examples of the regioselectivity of electrophilic addition.

major   minor
product   product

major   minor
product   product

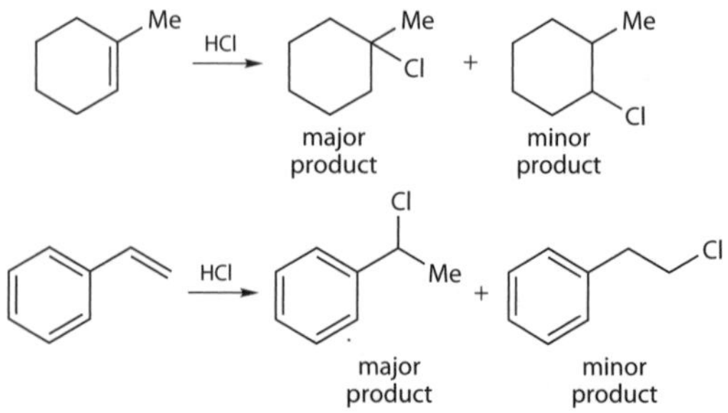

Carbocation stability

■ The regioselectivity of the electrophilic addition is due to the reaction proceeding through a carbocation intermediate. It is the stability of this intermediate that will determine which addition product is formed as the major product.

■ A carbocation bears a formal positive charge; it is an electron-deficient species. As such it will be stabilised through the inductive donation of electron density from adjacent groups.

■ The more highly substituted a carbocation, the greater the stability of the carbocation (Figure 6.17).

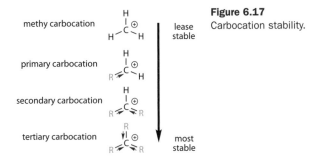

Figure 6.17
Carbocation stability.

■ More highly substituted carbocation intermediates are favoured in reactions owing to their lower energy compared with less substituted carbocations. A reaction will proceed predominantly via the pathway that involves the lowest-energy intermediate.

■ To determine the major product of an electrophilic addition reaction it is essential to consider carbocation stability:

tertiary carbocation > secondary carbocation > primary carbocation

Figure 6.18
Determining the major product of addition (carbocation stability).

■ The first step of the electrophilic addition in Figure 6.18 is the addition of H^+ to the double bond.

■ Since the double bond is not symmetric, the initial addition of H^+ at either of the two ends will give two non-equivalent carbocations (Figure 6.19).

■ The more highly substituted carbocation will be lowest in energy and lead to the major product observed experimentally for the reaction.

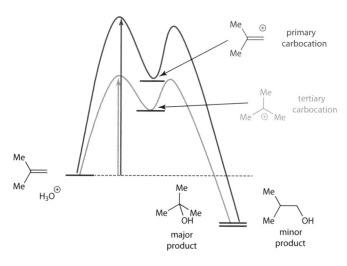

Figure 6.19
Reaction profile for addition to non-symmetric double bonds.

Carbocation stability and resonance

- The more stable carbocation will always lead to the major product of a reaction (Figures 6.20 and 6.21).
- The carbocation leading to the formation of product 1 is a secondary carbocation (Figure 6.22).
- The carbocation leading to the formation of product 2 is also a secondary carbocation. In this case though the carbocation is directly adjacent to the aromatic ring (Figure 6.23).
- The close proximity of the aromatic ring enables the carbocation to be delocalised into ring through the overlap of the empty p orbital of the carbocation with the filled p orbitals of the ring.
- This delocalisation of the positive charge provides an additional resonance stabilisation for carbocation 2.

Figure 6.20
Determining the major product of addition (resonance).

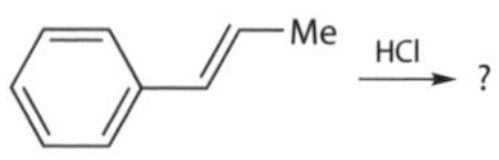

Figure 6.21
Electrophilic addition of HCl to a conjugated alkene.

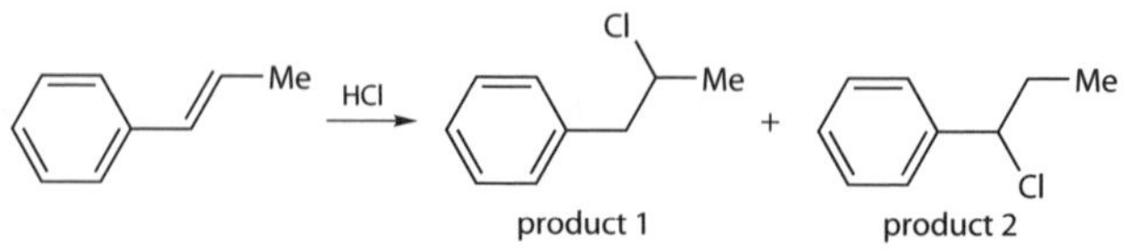

product 1

product 2

Figure 6.22
Formation of addition product 1.

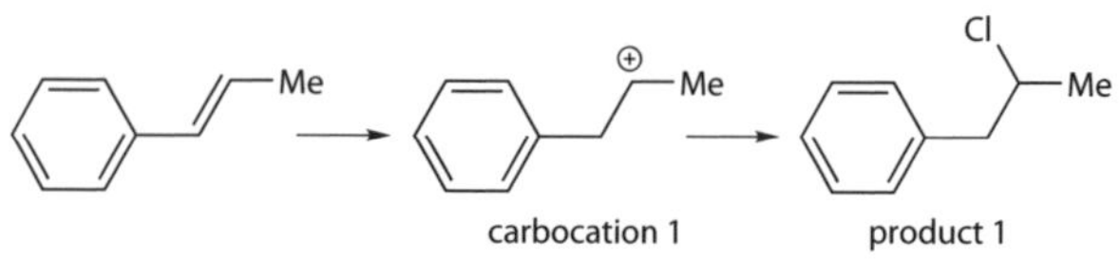

carbocation 1

product 1

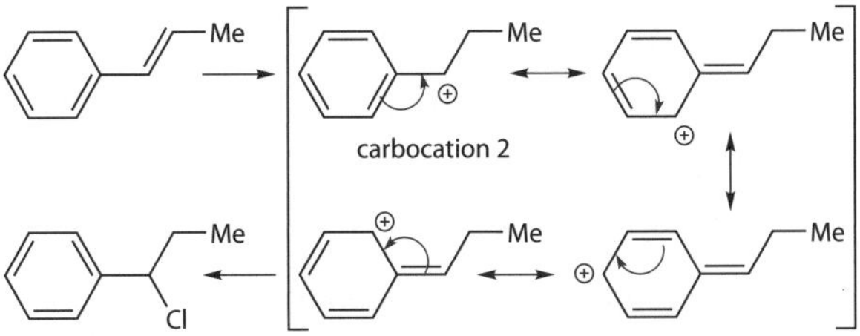

Figure 6.23
Formation of addition product 2.

carbocation 2

product 2

■ Carbocation 1 cannot stabilise the cation by resonance and so is higher in energy than carbocation 2 despite both being secondary carbocations. The major product of the reaction would therefore be product 2.

Nucleophilic substitution

■ Alkyl halides are susceptible to nucleophilic attack due to the inductive withdrawal of electron density by the halogen atom which creates an electron-poor, electrophilic carbon (Figure 6.24).

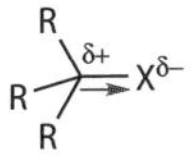

Figure 6.24
Electrophilic character of alkyl halides.

■ When a nucleophile attacks an electrophilic sp^3 hybridised alkyl halide, a new bond is created between the nucleophile and the electron-poor carbon with the halide being displaced in the process. This sequence of events is not always simultaneous.

KeyPoints

■ There are two classes of nucleophilic substitution reaction, S_N1 and S_N2.

S_N1			S_N2		
substitution		unimolecular	substitution		bimolecular
	nucleophilic			nucleophilic	

■ The distinction between these two classes lies in the timing of the bond-breaking and making processes and the chemical species involved in those processes.
■ A unimolecular process is one in which there is only *one* chemical species involved in the rate-determining step of the reaction.
■ A bimolecular process is one in which there are *two* chemical species involved in the rate-determining step of the reaction.

S$_N$1 reactions

- In the S$_N$1 reaction the breaking of the carbon–halogen bond is more advanced than the formation of the nucleophile–carbon bond (Figure 6.25). This results in the loss of the halogen to form a carbocation prior to nucleophilic attack.
- The nucleophile attacks the carbocation in the second step of the reaction.
- The rate-determining step of the reaction is the formation of the carbocation; this step is unimolecular as only the alkyl halide is involved.

$$Rate = k \text{ [alkyl halide]}$$

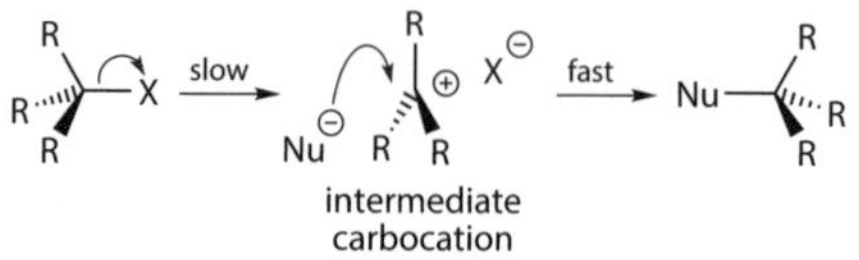

Figure 6.25
Generalised mechanism for S$_N$1 reactions.

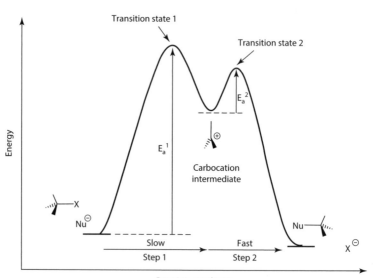

intermediate carbocation

Figure 6.26
Reaction profile for S$_N$1 reactions.

Changes in hybridisation during S$_N$1 reactions

- Initially, the carbon of the alkyl halide is sp^3 hybridised. Upon breaking of the carbon–halogen bond the pair of electrons from this bond localise on the halogen, leaving an electron-deficient carbon with a share of only six electrons.
- The hybridisation of the carbon atom changes to one of sp^2 hybridisation to minimise the repulsion of the bonding pairs of electrons, leaving the p orbital empty (Figure 6.27).

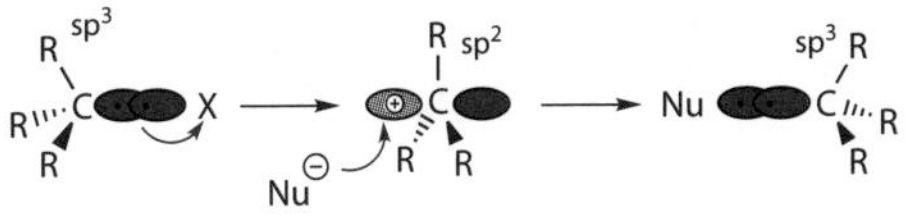

Figure 6.27
Changes in hybridisation.

- The nucleophilic Nu⁻ is able to attack the planar sp² hybridised carbocation from either side of the carbon framework, donating a pair of electrons into the empty p orbital.
- To form a bond to the nucleophile the hybridisation of the central carbon must change to one of sp³ hybridisation. There has been no net change in hybridisation for the reaction, despite a change in hybridisation for the intermediate carbocation.

Stereochemistry in S$_N$1 reactions

- One implication of having a planar sp² hybridised carbocation intermediate is that when the carbocation bears three different substituents, addition of the nucleophile to either the left face (arrow a in Figure 6.28) or the right face (arrow b) of the carbocation will not yield the same product.
- The two molecules formed are mirror images of each other. They are enantiomers and, since the nucleophile has an equal probability of adding to the left face as it does to the right face, the two enantiomers will be formed in equal amounts, and a racemic mixture will result.
- If the starting molecule bears four different substituents, it is chiral; the S$_N$1 reaction in going through a planar carbocation intermediate that destroys this chirality and racemises the molecule.
- Two enantiomers of the same starting molecule would result in the same racemic mixture being formed since both enantiomers would go through the same carbocation intermediate.

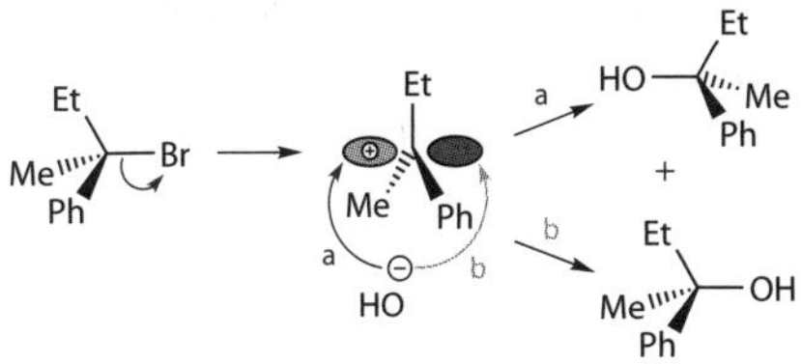

Figure 6.28
Nucleophilic attack at the sp² hybridised carbon.

Factors affecting S$_N$1 reactions

Nature of the substrate

- The reactivity of a substrate towards an S$_N$1 reaction is determined by the stability of the intermediate carbocation (Figure 6.29). Carbocation stability decreases in the order:

tertiary carbocation > secondary carbocation > primary

carbocation > methyl carbocation

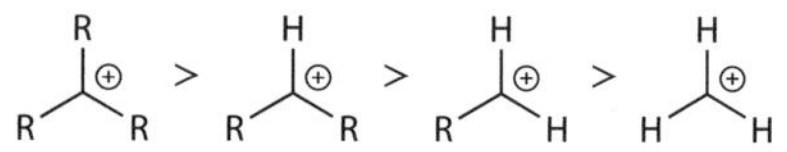

Figure 6.29
Carbocation stability.

- As carbocation stability increases, reactivity towards an S_N1 reaction also increases.
- Carbocations will be stabilised by substituents that donate electron density by induction and by groups that donate electron density by resonance.

Nature of the leaving group

- The rate of formation of the intermediate carbocation will also be affected by the stability of the leaving group. The more stable the anion being formed, the better the leaving group will be. A good leaving group will therefore be a weak base, that is, the conjugate base of a strong acid (Figure 6.30).

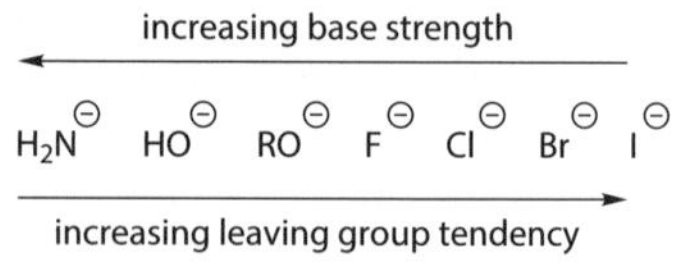

Figure 6.30
Leaving group tendencies.

Nature of the solvent

- Although the starting material and product are of comparable polarity, the carbocation intermediate is considerably more polar than the starting material.
- A polar solvent will provide a greater stabilising effect on the intermediate through the increased solvation of the charged carbocation.
- In stabilising the carbocation, the rate of the reaction will be increased since the intermediate will be of lower energy compared to when a non-polar solvent is used.
- Suitable polar solvents for the S_N1 reaction include water, methanol and formic acid.

Nature of the nucleophile

- The nature of the nucleophile will not influence the S_N1 reaction since the nucleophile is not involved in the rate-determining step.

Rearrangements in S_N1 reactions

- The carbocation intermediate will rearrange where possible to form a more highly substituted and hence more stable carbocation.
- Carbocation rearrangement will occur through either a hydride shift or a methyl shift (Figure 6.31).

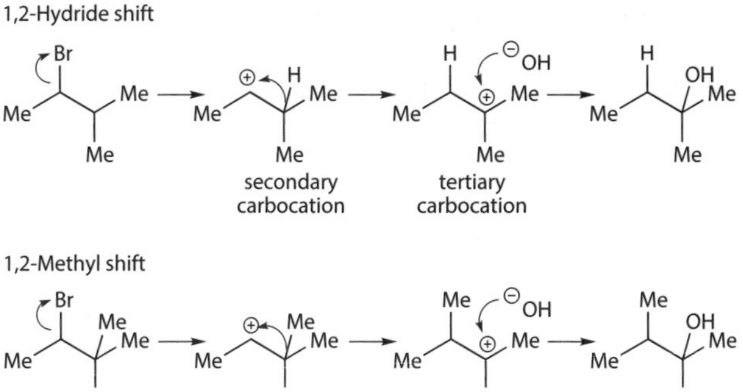

1,2-Hydride shift

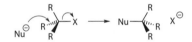

secondary
carbocation

tertiary
carbocation

Figure 6.31
Carbocation rearrangements.

1,2-Methyl shift

S$_N$2 reactions

- In the S$_N$2 reaction the breaking of the carbon–halogen bond and the formation of the nucleophile–carbon bond occur simultaneously (Figure 6.32).
- The nucleophile attacks and the leaving group is displaced in a single step.
- The rate-determining step of the reaction is the formation of the substituted product; this step is bimolecular as both the nucleophile and the alkyl halide are involved.

- Primary carbocations will rearrange to the more stable secondary or tertiary carbocation.
- Secondary carbocations will rearrange to the more stable tertiary carbocation.
- Tertiary carbocations will never rearrange.

Rate = k [alkyl halide][nucleophile]

Figure 6.32
Generalised mechanism for S$_N$2 reactions.

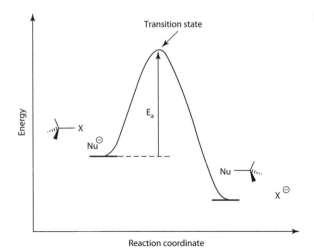

Figure 6.33
Reaction profile for S$_N$2 reactions.

Transition state

E_a

Energy

Reaction coordinate

Changes in hybridisation during the S_N2 reaction

- The incoming nucleophile is attacking at the same time as the leaving group is leaving; there is no formal change in the hybridisation of the molecule during the reaction (Figure 6.34).
- The transition state for the S_N2 reaction is pentacoordinate with the pair of electrons from the nucleophile entering the σ* orbital of the carbon–halogen bond.
- Although the transition state may appear as though the central carbon is sp² hybridised, it is not.

Figure 6.34
Changes in hybridisation.

Stereochemistry in S_N2 reactions

- Since the S_N2 reaction is concerted, the nucleophile must attack from the side of the molecule opposite the leaving group (Figure 6.35).
- If the starting molecule bears four different substituents, it is chiral; the S_N2 reaction will yield a product in which the stereochemistry of the central carbon atom has been inverted in much the same way as an umbrella turning inside out in strong winds.

Figure 6.35
Nucleophilic attack at the sp³ hybridised carbon.

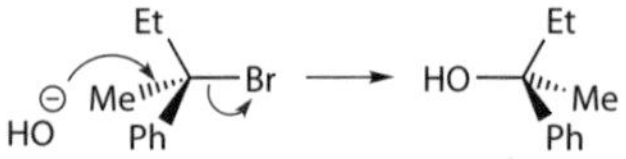

Factors affecting S_N2 reactions

Nature of the substrate

- The reactivity of a substrate towards an S_N2 reaction is highly dependent on the steric environment around the electrophilic carbon atom (Figure 6.36).

Figure 6.36
Steric crowding in the substrate.

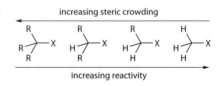

- The more highly substituted the carbon, the less reactive the substrate is towards an S_N2 reaction since it is more difficult for the nucleophile to approach the central carbon atom.

Nature of the leaving group

■ The role of the leaving group in an S_N2 reaction will be the same as in an S_N1 reaction (Figure 6.37). The more stable the anion being formed, the better the leaving group will be. A good leaving group will therefore be a weak base, that is, the conjugate base of a strong acid.

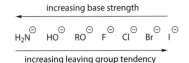

Figure 6.37
Leaving group tendencies.

Nature of the solvent

■ Unlike the S_N1 reaction where the solvent can stabilise the intermediate carbocation, in an S_N2 reaction it is the interaction between the nucleophile and the solvent that is important.
■ The nucleophile is an electron-rich species, and will be most soluble in polar solvents.
■ It is important that the solvent is aprotic (without an acidic proton) since protic solvents are able to bond hydrogen to the nucleophile and would create an extended solvent cluster that would hinder the approach of the nucleophile to the electrophilic carbon.
■ Suitable polar aprotic solvents for the S_N2 reaction include dimethyl sulfoxide (DMSO), N,N-dimethylformamide (DMF), tetrahydrofuran (THF), acetonitrile and acetone.

Nature of the nucleophile

■ Since the nucleophile is involved in the rate-determining step for the S_N2 reaction, the more nucleophilic this species is, the faster the reaction will be.
■ Rules for determining nucleophilicity:
   1. For groups containing the same atom, anions will be more nucleophilic than neutral species.

$$HO^{\ominus} > H_2O \qquad RO^{\ominus} > ROH$$

   2. As you go down a group in the periodic table, nucleophilicity increases. The electrons are less tightly held by the nucleus as there are additional shells of electrons shielding the positive charge of the nucleus, making the outermost electrons more easily polarised.

$$I^{\ominus} > Br^{\ominus} > Cl^{\ominus} > F^{\ominus}$$

3. As you go right across the periodic table, nucleophilicity decreases. The electrons are being added to the same orbitals but the number of protons in the nucleus is increasing. Electrons in atoms further to the right experience a greater effective nuclear charge which makes them less available for donation.

$$HO^{\ominus} > F^{\ominus} \qquad NH_3 > H_2O$$

4. Cyclic amines are more nucleophilic than acyclic amines. For cyclic amines the ring ties the alkyl groups back, resulting in a more directional lone pair of electrons on the nitrogen. For acyclic amines the alkyl groups are constantly in motion, as is the lone pair of electrons.

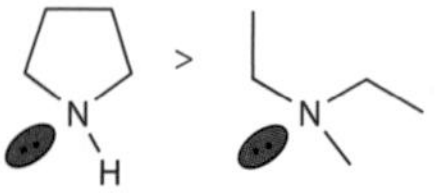

Elimination reactions

■ Elimination reactions can be considered to be the opposite of electrophilic addition reactions. Instead of adding HX to a double bond, H and X are being removed from adjacent carbon atoms to form a new double bond (Figure 6.38).

Figure 6.38
Generalised mechanism for elimination.

■ Nucleophiles are able to displace a leaving group from a saturated carbon centre through the donation of electron density into the substrate in a substitution reaction. However, the lone pair of electrons on the nucleophile will also give the nucleophile some basic character.

■ Elimination reactions occur when a nucleophile is sufficiently basic that, instead of attacking at carbon, it acts as a base and deprotonates adjacent to a leaving group (Figure 6.39). The electron density from the carbon–hydrogen bond forms a new carbon–carbon double bond and displaces the leaving group in the process. The sequence of events is not always simultaneous.

Figure 6.39
Regioselectivity in elimination reactions.

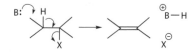

major product   minor product

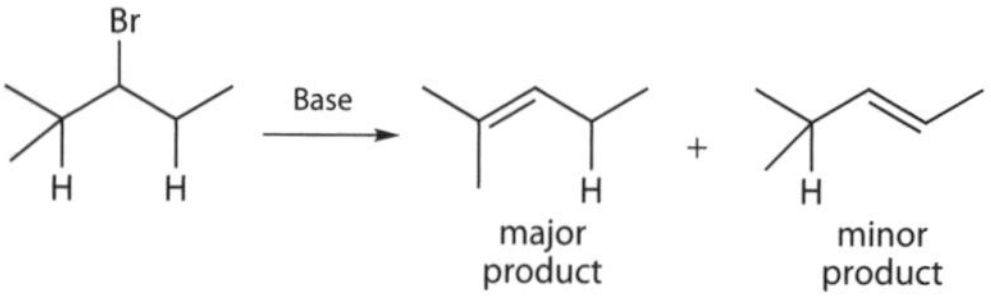

Zaitsev's rule

- When removing HX or H_2O to form a double bond, the hydrogen that is lost will come from the carbon having the fewest hydrogen substituents. This will give the most highly substituted product.

KeyPoints

- There are two classes of elimination reaction, E1 and E2.

- The distinction between these two classes lies in the timing of the bond-breaking and making processes and the chemical species involved in those processes.

E1 reactions

- In the E1 reaction the breaking of the carbon–halogen bond is more advanced than deprotonation and the formation of the carbon–carbon bond. This results in the loss of the halogen to form a carbocation intermediate prior to deprotonation (Figure 6.40).

Tip

Elimination reactions share similarities with both electrophilic addition and nucleophilic substitution reactions.

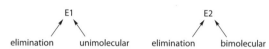

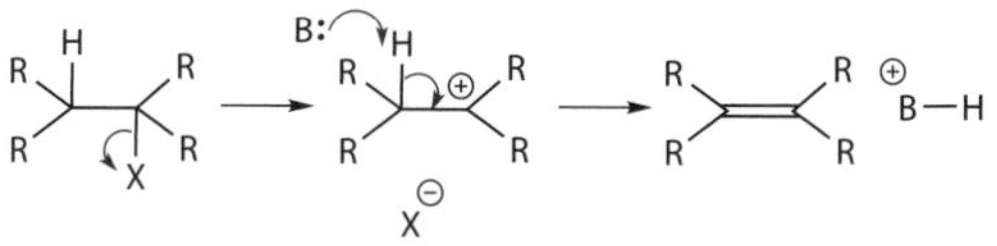

Figure 6.40
Generalised mechanism for E1 reactions.

- The nucleophile deprotonates adjacent to the carbocation in the second step of the reaction to form the alkene product, quenching the positive charge in the process.
- The rate-determining step of the reaction is the formation of the carbocation; this step is unimolecular as only the alkyl halide is involved.

$$\text{Rate} = k\,[\text{alkyl halide}]$$

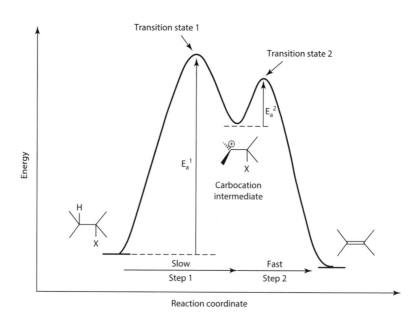

Figure 6.41 Reaction profile for E1 reactions.

Changes in hybridisation during E1 reactions

- Initially, the carbon of the alkyl halide is sp^3 hybridised. Upon breaking of the carbon–halogen bond the pair of electrons from this bond localise on the halogen, leaving an electron-deficient carbon with a share of only six electrons.
- The hybridisation of the carbon atom changes to one of sp^2 hybridisation to minimise the repulsion of the bonding pairs of electrons, leaving the p orbital empty.
- The nucleophile, behaving as a base, attacks the hydrogen on the adjacent carbon atom, with the electrons from the carbon–hydrogen bond being donated to the vacant p orbital of the carbocation, forming a π bond.
- To form the π bond the carbon–hydrogen bond and the vacant p orbital must be in the same plane. The hybridisation of the carbon must change to one of sp^2 hybridisation.
- The two sp^3 hybridised carbons in the starting alkyl halide have both been converted to sp^2 hybridised carbons during the course of the reaction (Figure 6.42).

Figure 6.42
Changes in hybridisation.

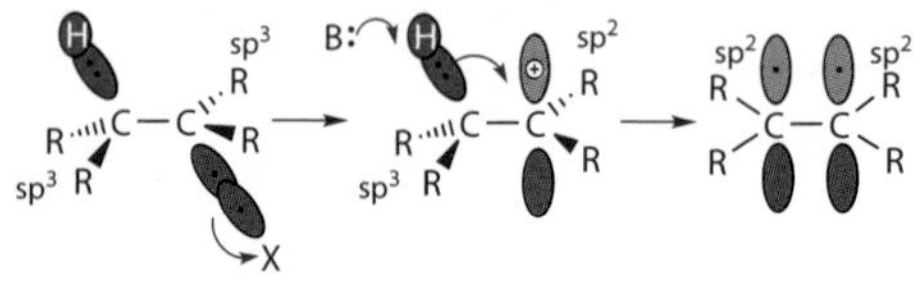

Stereochemistry in E1 reactions

- There is free rotation about the carbon–carbon bond of the carbocation intermediate in an E1 reaction. This rotation allows two conformers to be adopted which fulfil the requirement of having the carbon–hydrogen bond in the same plane as the vacant p orbital of the carbocation (Figure 6.43).
- These two conformers will not be of equal energy.
- The Ph and Me substituents of conformer b are located in close proximity. The steric hindrance experienced will disfavour this conformation in favour of *conformer a*, where these groups are located opposite each other. The major product will therefore be the *E*-alkene.

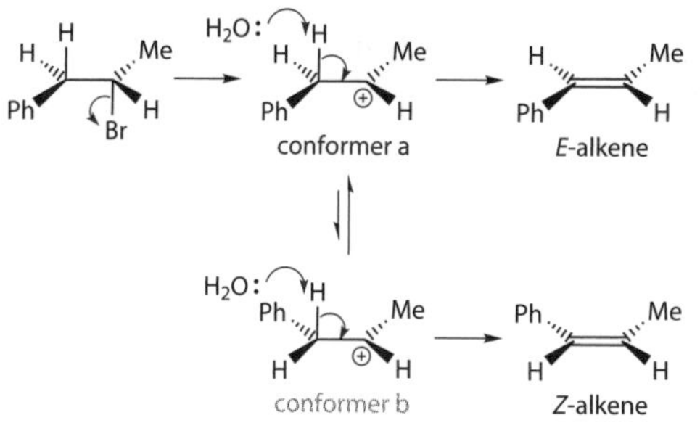

Figure 6.43
Conformer stability determines major product formation.

Factors affecting E1 reactions

Nature of the substrate

- The reactivity of a substrate towards an E1 reaction is determined by the stability of the intermediate carbocation. Carbocation stability decreases in the order:

tertiary carbocation > secondary carbocation > primary carbocation

- As carbocation stability increases, reactivity towards an E1 reaction also increases.
- Carbocations will be stabilised by substituents that donate electron density by induction and substituents that donate electron density by resonance.
- Tertiary alkyl halides will undergo an E1 reaction, as will some secondary alkyl halides. Primary alkyl halides will never undergo an E1 reaction as the intermediate carbocation would be too unstable.

Nature of the leaving group

- The rate of formation of the intermediate carbocation will be affected by the stability of the leaving group. The more stable the anion being formed, the better the leaving group will be.

Nature of the solvent

■ The carbocation intermediate will be stabilised by a polar solvent.
■ Suitable polar solvents for the E1 reaction include water, methanol and formic acid.

Nature of the base

■ The nature of the base/nucleophile will not influence the E1 reaction since the base is not involved in the rate-determining step. The reaction will proceed the same in the presence of either a strong base or a weak base.

E2 reactions

■ In the E2 reaction the deprotonation, formation of the π bond and breaking of the carbon–halogen bond occur simultaneously.
■ The base attacks and the leaving group is displaced in a single step (Figure 6.44).
■ The rate-determining step of the reaction is the formation of the eliminated product; this step is bimolecular as both the base and the alkyl halide are involved.

$$\text{Rate} = k \text{ [alkyl halide][base]}$$

Figure 6.44
Generalised mechanism for E2 reactions.

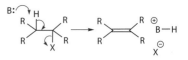

Figure 6.45
Reaction profile for E2 reactions.

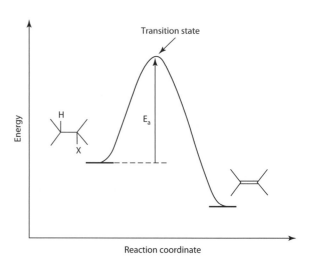

Changes in hybridisation during E2 reactions

■ The two adjacent carbon centres of the alkyl halide participating in the elimination must both undergo a change in hybridisation from sp^3 to sp^2 hybridisation to form the π bond of the alkene product (Figure 6.46).

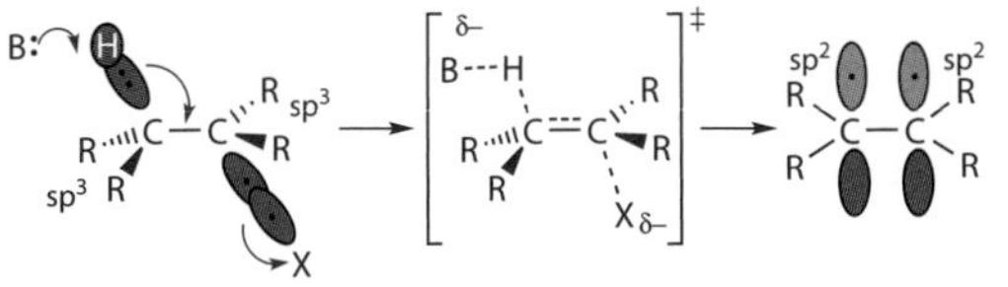

Figure 6.46
Changes in hybridisation.

Stereochemistry in E2 reactions

■ Since the E2 reaction is concerted, the carbon–hydrogen bond and the carbon–halogen bond must be in the correct relative position prior to deprotonation by the base.
■ The two bonds need to be in the same plane and opposite each other for the reaction to occur. That is, they should have an antiperiplanar conformation (Figure 6.47).

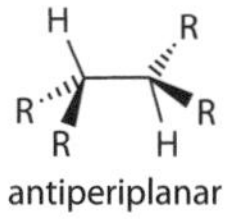

Figure 6.47
Antiperiplanar conformation of substituents for E2 reaction.

antiperiplanar

■ This conformation ensures that the σ bonds being broken are of the correct orientation to form the π bond of the product.
■ Where two antiperiplanar conformers are possible, the major product will be formed from the lowest-energy transition state in which steric repulsion between adjacent groups has been minimised (Figure 6.48).

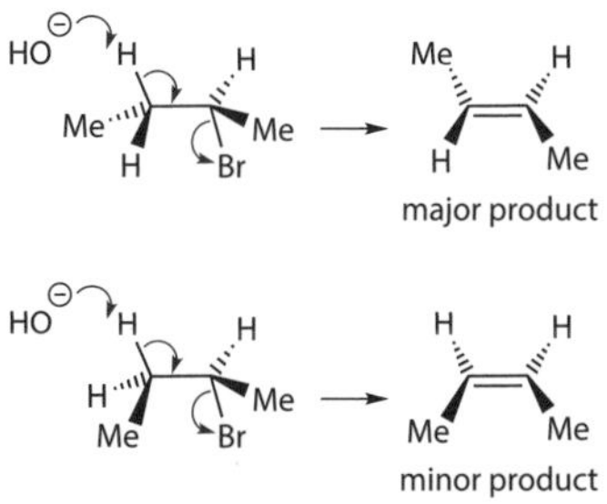

Figure 6.48
Conformer stability determines major product formation.

major product

minor product

Factors affecting E2 reactions

Nature of the substrate
- The presence of electron-donating alkyl substituents on the carbon bearing the halide will facilitate the breaking of this bond.
- Tertiary alkyl halides will therefore react faster than secondary or primary alkyl halides (Figure 6.49).

Figure 6.49
Reactivity of alkyl halides towards elimination.

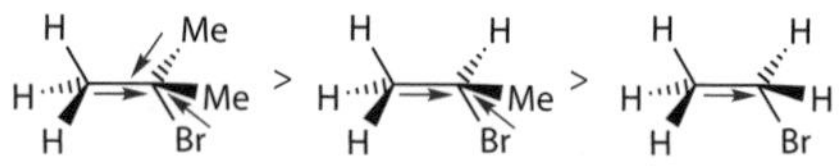

Nature of the leaving group
- The carbon–halide bond is broken in the rate-determining step. Therefore the better the leaving group, the faster the reaction.
- The leaving group should not be so good as to favour the E1 reaction.

Nature of the solvent
- Polar aprotic solvents will favour the E2 reaction, solvating both the transition state and the base without forming hydrogen-bonding interactions to the base that would hinder approach to the substrate.
- Suitable solvents for the E2 reaction include DMSO, DMF and acetonitrile.

Nature of the base
- The base is present in the rate equation for the reaction; therefore, as the strength of the base increases, so too does the rate of the reaction.
- Strong, negatively charged bases such as hydroxide, methoxide and t-butoxide promote the E2 reaction.
- Hindered bases will react at the least hindered hydrogen–carbon bond (blue bond) to give the least substituted product (Figure 6.50).

Figure 6.50
Changing the base changes the selectivity of E2 reactions.

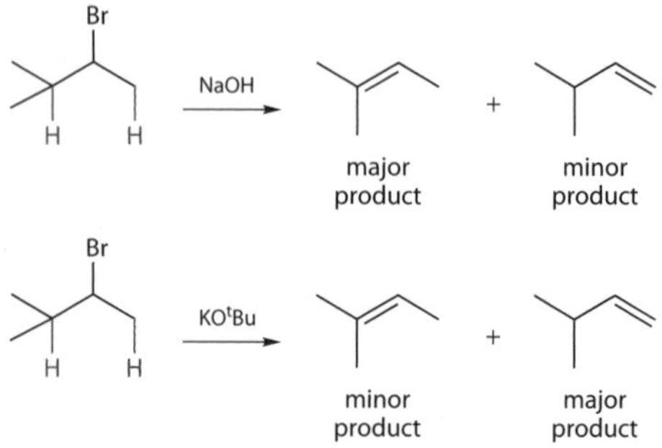

Competition between substitution and elimination

▪ Good nucleophiles that are weak bases favour substitution reactions over elimination reactions.

▪ Bulky, non-nucleophilic bases favour elimination reactions over substitution reactions.

▪ Tertiary alkyl halides can undergo S_N1, E1 and E2 reactions but not S_N2 reactions owing to steric crowding in the transition state.

▪ Secondary alkyl halides can undergo S_N1, S_N2, E1 and E2 reactions.

▪ Primary alkyl halides can undergo S_N2 and E2 reactions which are concerted and do not involve the formation of an unstable carbocation intermediate.

Self-assessment

1. Which of the following alkyl bromides is most reactive towards an S_N1 reaction?

a.

b.

c.

d.

2. In the electrophilic bromination of alkenes, which one of the following intermediates can be used to explain the stereochemical outcome of the addition?

a.

b.

c.

d.

3. **Which of the following rate equations correctly describes the E2 reaction of alkyl halides?**
a. rate = k [alkyl halide]
b. rate = k [base]
c. rate = k [alkyl halide][base]
d. rate = k [alkyl halide][base]^2

4. **Which of the following is the most stable carbocation?**

a.
$$Ph-\overset{\overset{\displaystyle H}{|}}{C}{}^{\oplus}-Ph$$

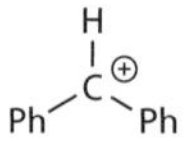

b.
$$Ph-\overset{\oplus}{C}H_2$$

c.
$$H_3C-\overset{\overset{\displaystyle H}{|}}{C}{}^{\oplus}-Ph$$

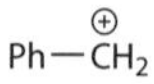

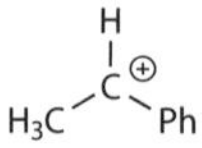

d.
$$H-\overset{\overset{\displaystyle H}{|}}{C}{}^{\oplus}-H$$

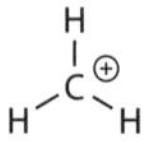

5. **What would be the major product of the following reaction?**

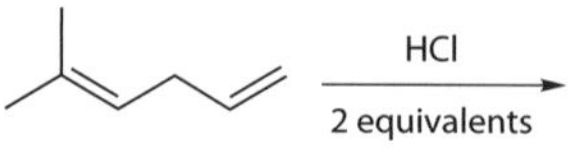

HCl
2 equivalents

a.

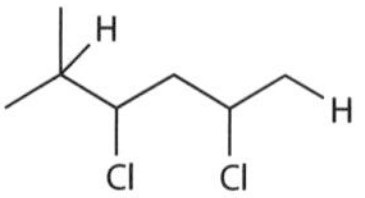

b.

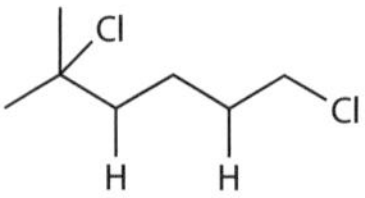

c.

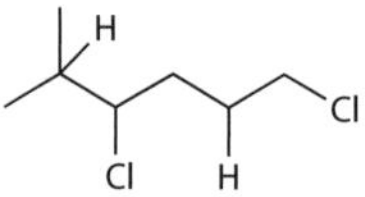

d.

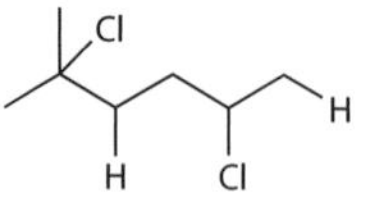

6. **For a regioselective reaction, which of the following statements is true?**
a. Both possible products are formed in equal amounts as a 1:1 mixture.
b. One of the possible products is formed in larger amounts than the other product.
c. Only one of the possible products is formed.
d. Depending on reaction conditions used, all of the above statements can be true.

7. **Which of the following solvents would be most suitable for an S$_N$2 reaction?**
a. acetonitrile
b. dichloromethane
c. diethyl ether
d. methanol

8. **A tertiary carbocation is more stable than either a secondary or primary carbocation because:**
a. it is sp^2 hybridised
b. it posesses one electron-donating group
c. it possesses three electron-withdrawing groups
d. it possesses three electron-donating groups

9. **The major product(s) of an E1 reaction will be:**
a. an *E*-alkene
b. a *Z*-alkene
c. a *meso* compound
d. a 50:50 mixture of *E*- and *Z*-alkenes

10. **When an S$_N$2 reaction occurs with a single stereoisomer of a chiral compound the resulting product is:**
a. achiral
b. chiral
c. a racemic mixture
d. optically inactive

chapter 7

Chemistry of aromatic compounds

Overview

After learning the material presented in this chapter you should:

- understand the stabilisation granted by aromaticity
- understand the reactivity of the aromatic ring towards electrophilic substitution
- appreciate the need for a powerful electrophile in these reactions and understand how they are formed
- be able to determine the effect of electron-donating and electron-withdrawing groups on reactivity
- appreciate the properties of monosubstituted and disubstituted benzene derivatives and understand their reactivity patterns.

Benzene

- All carbon–carbon bonds in benzene are identical; the six-membered ring is completely symmetric.
- The six carbon–carbon bonds have the same length (13.9 nm), intermediate between a single bond (15.4 nm) and a double bond (13.4 nm).
- In benzene each carbon atom is sp^2 hybridised with a single electron in a 2p orbital.
- The 2p orbitals overlap, delocalising electron density equally across the π system.
- The delocalisation of electron density provides benzene with additional resonance stabilisation when compared to a hypothetical alternating double bond–single bond system (Figure 7.1).

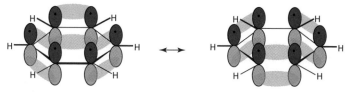

Figure 7.1
Delocalisation of electron density in benzene.

Alternating double and single bonds

Resonance provides two configurations of bonds

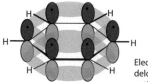

Electron density delocalised over entire ring system

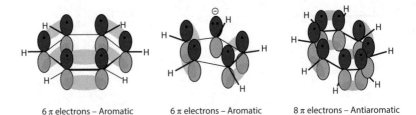

6 π electrons – Aromatic   6 π electrons – Aromatic   8 π electrons – Antiaromatic

Figure 7.2 Aromatic compounds obey Hückel's rule.

- The additional stabilisation afforded by the overlapping of adjacent p orbitals in benzene is termed aromaticity.
- For a molecule to be aromatic there needs to be delocalisation of electron density through adjacent p orbitals in a planar, cyclic system with 4n+2 π electrons within that system (Hückel's rule: Figure 7.2).

Naming of benzene derivatives

- The positions on a benzene ring relative to a functional group are given as *ipso*, *ortho*, *meta* and *para* (Figure 7.3). Where *ipso* refers to the carbon where the referencing functional group is located, the *ortho* position is adjacent to the referencing group, the *para* position is opposite the referencing group and the *meta* position is between the *ortho* and *para* positions.

Tip

The positions on the ring are relative to the substituent on the ring, not the way the molecule is drawn on the page!

- Some benzene derivatives have commonly used trivial names (Figure 7.4).

Figure 7.3
Naming of positions on the benzene ring.

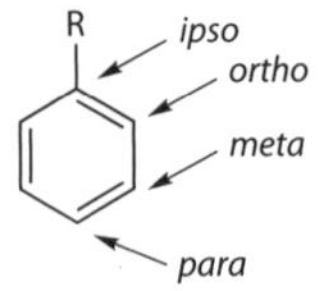

R — *ipso*
ortho
meta
para

Figure 7.4
Trivial names of some benzene derivatives.

CH₃   NH₂   OMe   OH   CO₂H

toluene   aniline   anisole   phenol   benzoic acid

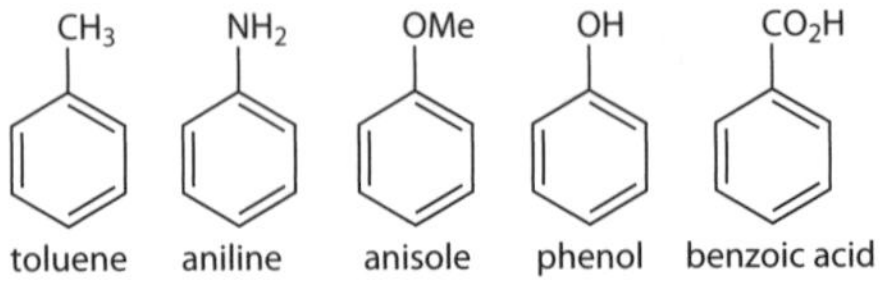

Reactivity

- Substituents attached to an aromatic ring are able to interact with the ring to either increase or decrease the electron density in the ring.

- Substituents which increase electron density in the ring are termed electron donating.
- Substituents which decrease electron density in the ring are termed electron withdrawing.
- There are two ways that the electron density of an aromatic ring can be either increased or decreased:
  1. induction
  2. resonance.
- The inductive effect operates through σ bonds and is the push/pull of electron density resulting from the difference in electronegativity between the two atoms of the bond.
- The electron density of a bond is not evenly distributed; the more electronegative atom will withdraw electron density towards itself (Figures 7.5 and 7.6).

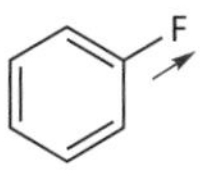

Figure 7.5
Inductively electron-withdrawing groups.

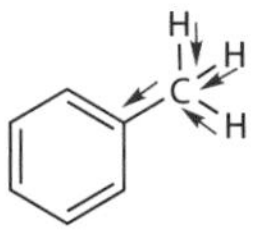

Figure 7.6
Inductively electron-donating groups.

- Resonance or the mesomeric effect operates through π bonds and is the push/pull of electron density through a conjugated system of overlapping p orbitals.
- An atom adjacent to an aromatic ring with a non-bonding pair of electrons will be able to push this pair of electrons into the ring; the substituent is electron donating by resonance (Figure 7.7).

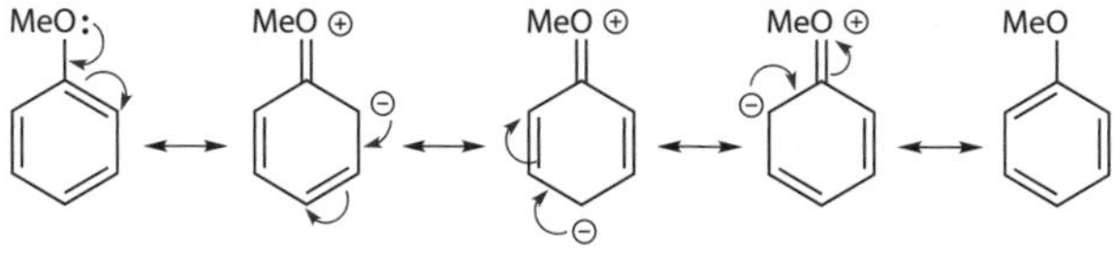

Figure 7.7
Resonance with an electron-donating substituent.

- A group that is electron donating by resonance will increase electron density at the *ortho* and *para* positions. This increase in electron density increases the nucleophilicity of the ring, making it more reactive towards electrophiles than benzene. The ring is said to be activated.
- A group adjacent to an aromatic ring with a double or triple bond will be able to pull a pair of electrons out of the ring; the substituent is electron withdrawing by resonance (Figure 7.8).

Tips

Electron-donating groups = Activating
Electron-withdrawing groups = Deactivating

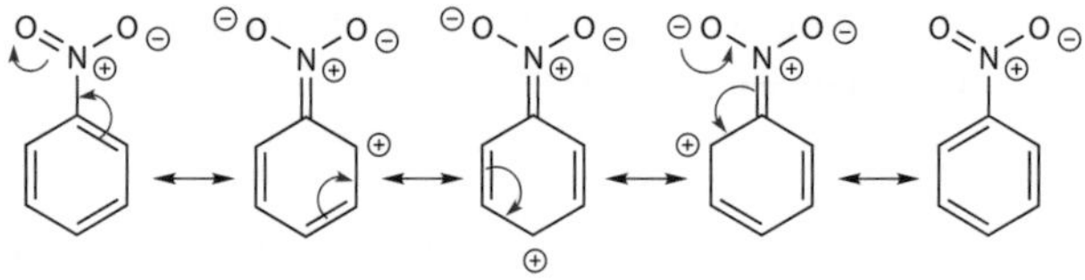

Figure 7.8
Resonance with an electron-
withdrawing substituent.

■ A group that is electron withdrawing by resonance will decrease
electron density at the *ortho* and *para* positions. This decrease
in electron density decreases the nucleophilicity of the ring,
making it less reactive towards electrophiles than benzene. The
ring is said to be deactivated.

■ When substituents are positioned with the correct orientation,
electron-donating groups and electron-withdrawing groups can
act synergistically (Figure 7.9). As one group pushes electron
density into the ring the other pulls the electron density out of
the ring.

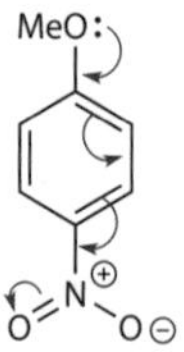

Figure 7.9
Synergy between electron-donating and
electron-withdrawing groups.

Electrophilic substitution

■ Whereas alkenes react readily with electrophiles in addition
reactions, the 'double bonds' of benzene by comparison are
relatively unreactive (Figure 7.10).

■ Benzene is a poor nucleophile compared with an isolated alkene;
this is due to benzene needing to interrupt the ring resonance to
donate a pair of electrons to an electrophile, resulting in a loss of
aromatic stabilisation.

■ A strong electrophile is required for benzene to react.

■ Benzene will not undergo addition reactions as this would result
in a permanent loss of aromaticity. Instead benzene undergoes
substitution reactions, where the loss of aromaticity is only
temporary and is restored in the product (Figure 7.11).

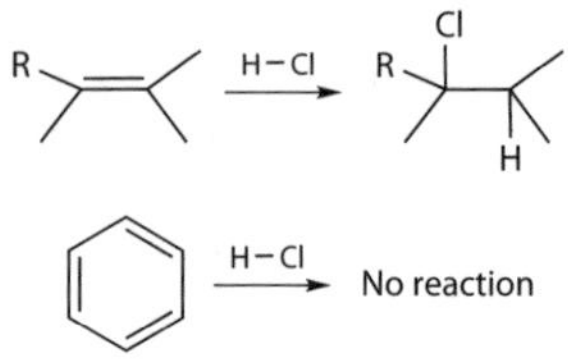

Figure 7.10
Reactivity of alkenes versus benzene.

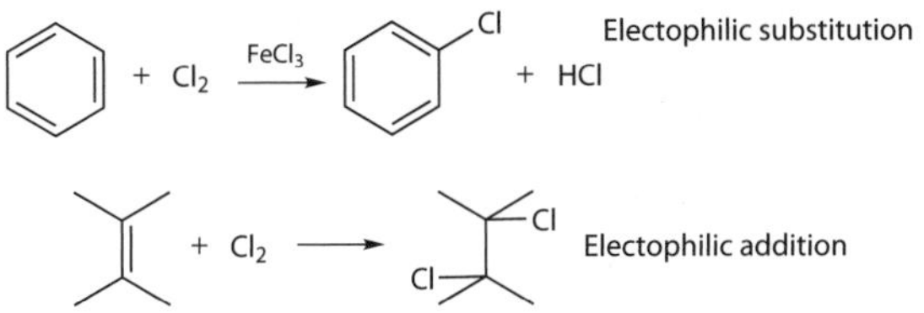

Figure 7.11
Electrophilic substitution and electrophilic addition.

Electrophilic substitution versus electrophilic addition

- When benzene reacts with a strong electrophile a pair of electrons from the ring system is lost to form an intermediate benzonium ion. This carbocation formation step is slow due to the loss of resonance stabilisation that accompanies the process.
- The benzonium ion can potentially undergo two different reactions. In the first, the counter ion of the electrophile deprotonates the carbon to which the electrophile added (arrow a in Figure 7.12), with the electrons from the carbon–hydrogen bond quenching the positive charge (arrow b), forming a new carbon–carbon double bond and restoring aromaticity. In the second possible reaction the counter ion directly attacks the carbocation (arrow c) to give the addition product. The addition product is not aromatic and had lost the stabilisation associated with a fully conjugated system. This product will not form.

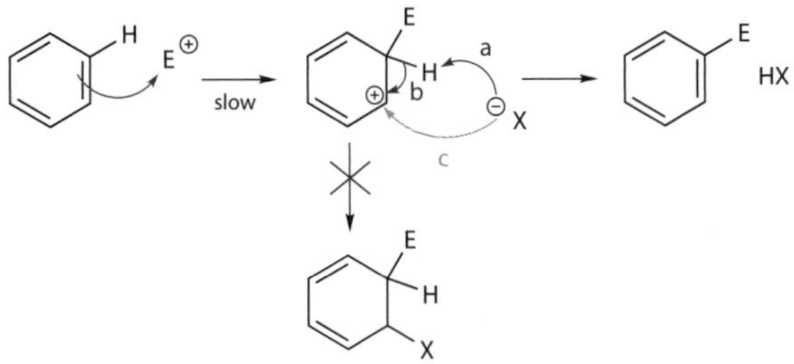

Figure 7.12
Generalised mechanism for electrophilic substitution.

Examples of electrophilic substitution

- The mechanism for all electrophilic substitution reactions begins with the formation of the powerful electrophile. Each electrophile then reacts identically to give the substituted benzene product (Figure 7.12).

Nitration of benzene

- Nitric acid and sulfuric acid react to form the highly electrophilic nitronium ion (Figure 7.15).

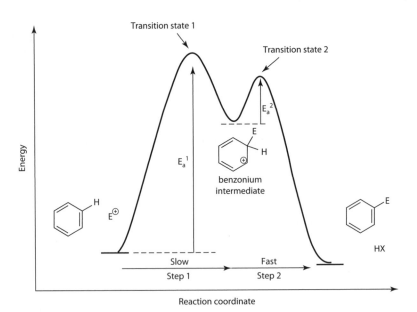

Figure 7.13 Reaction profile for electrophilic substitution.

Figure 7.14
Nitration of benzene.

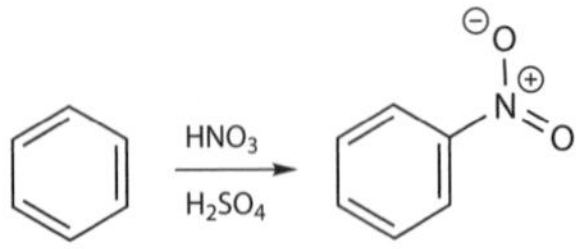

Figure 7.15
Formation of the electrophile.

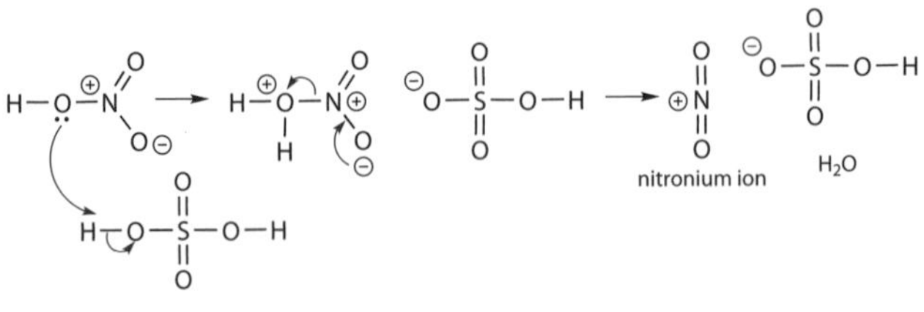

Figure 7.16
Reaction between
benzene and the
electrophile.

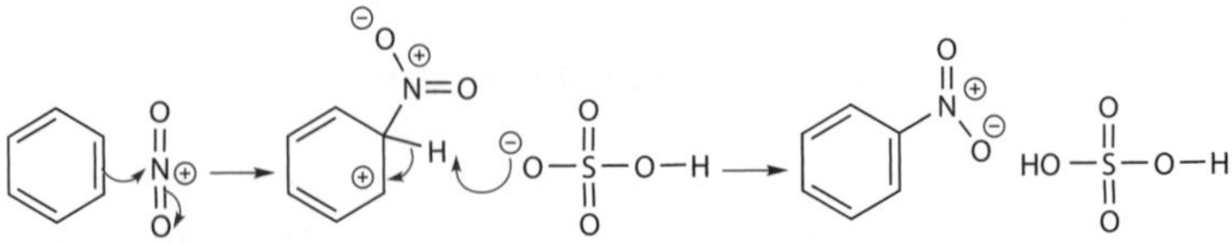

Sulfonation of benzene

■ The electrophile for sulfonation can be prepared in two ways,
either using fuming sulfuric acid (H_2SO_4 containing SO_3) or by
heating sulfuric acid (Figure 7.18).

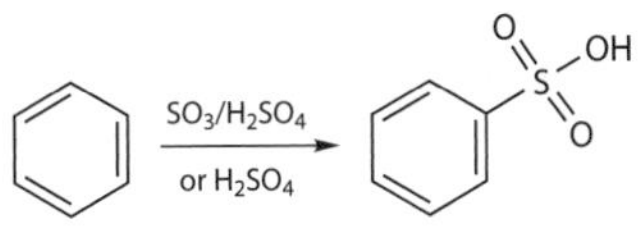

Figure 7.17
Sulfonation of benzene.

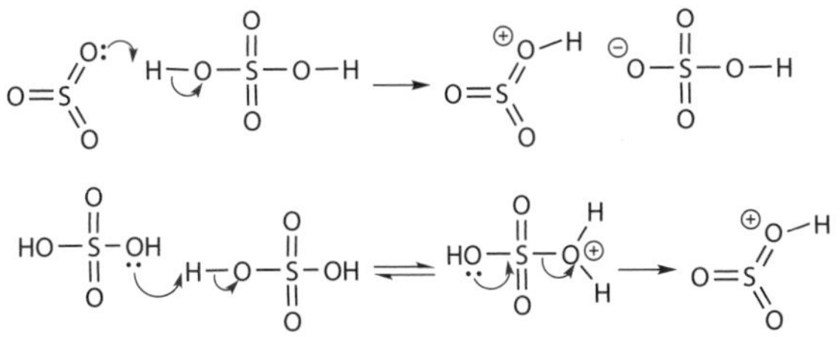

Figure 7.18
Formation of the electrophile.

- Whichever method is used to generate the protonated sulfur trioxide, the electrophile will react in the same way with benzene (Figure 7.19).
- Molecules have no memory of how they were formed and behave in the same way once formed.

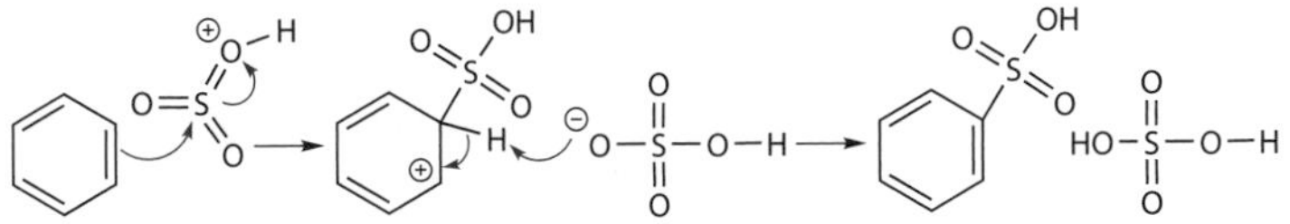

Figure 7.19
Reaction between benzene and the electrophile.

Halogenation of benzene

- AlCl$_3$ is a strong Lewis acid and is required to generate the highly reactive chloronium ion electrophile (Figure 7.21).

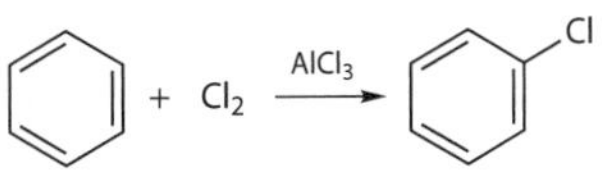

Figure 7.20
Halogenation of benzene.

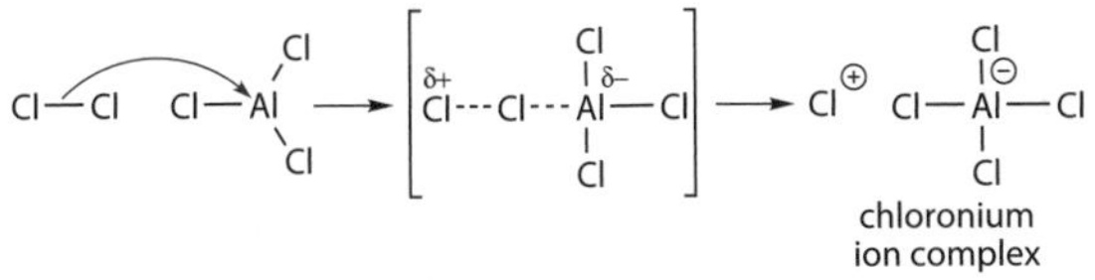

Figure 7.21
Formation of the electrophile.

chloronium
ion complex

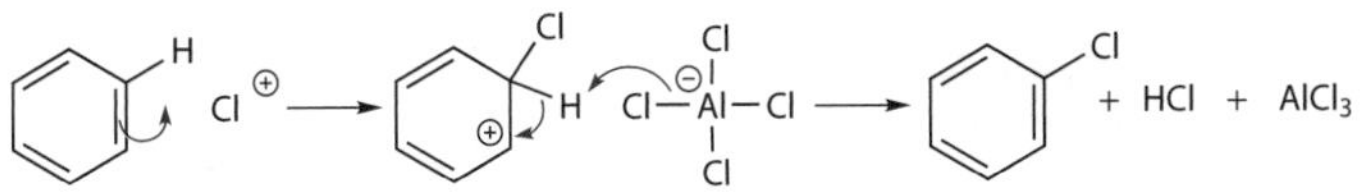

Figure 7.22
Reaction between benzene and the electrophile.

Friedel–Crafts alkylation

- AlCl$_3$ is a strong Lewis acid and is required to generate the highly reactive carbocation electrophile (Figure 7.24).
- Friedel–Crafts alkylations are prone to rearrangement of the intermediate carbocation.

Figure 7.23
Friedel–Crafts alkylation.

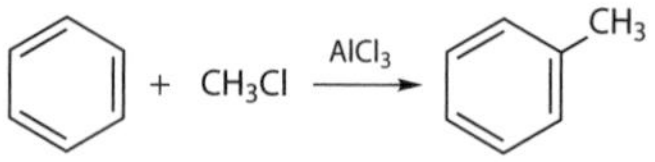

Figure 7.24
Formation of the electrophile.

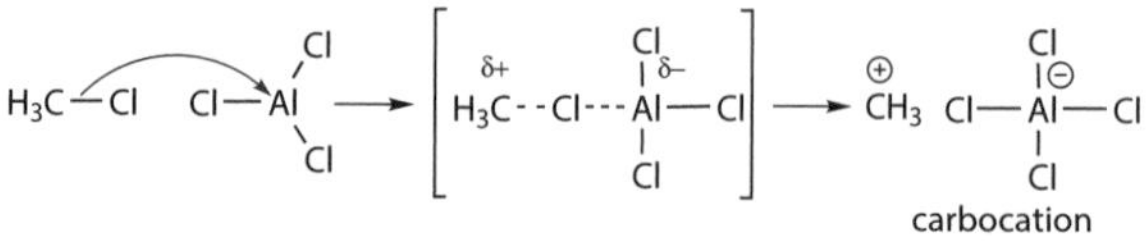

Figure 7.25
Reaction between benzene and the electrophile.

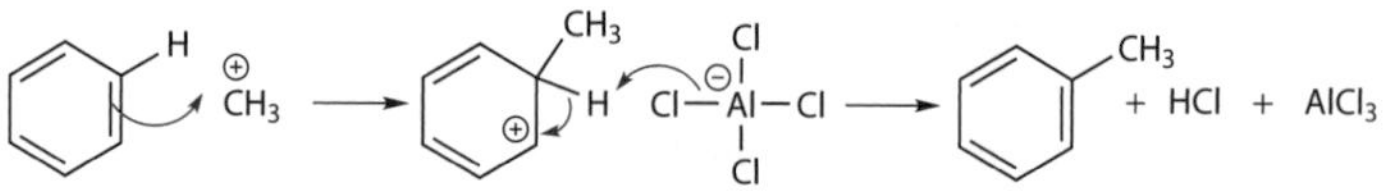

- Unstable primary carbocations undergo 1,2-hydride shifts to form more stable secondary or tertiary carbocations (Figure 7.26).
- The rearranged carbocation will then undergo electrophilic substitution with benzene (Figure 7.27).

Figure 7.26
Carbocation rearrangement in Friedel–Crafts alkylation.

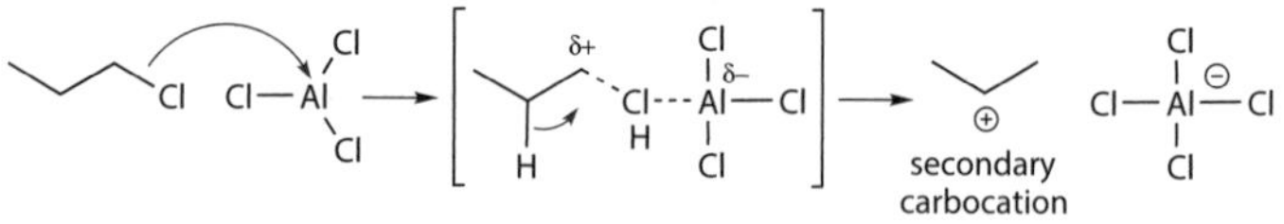

Figure 7.27
Reaction between benzene and rearranged carbocation.

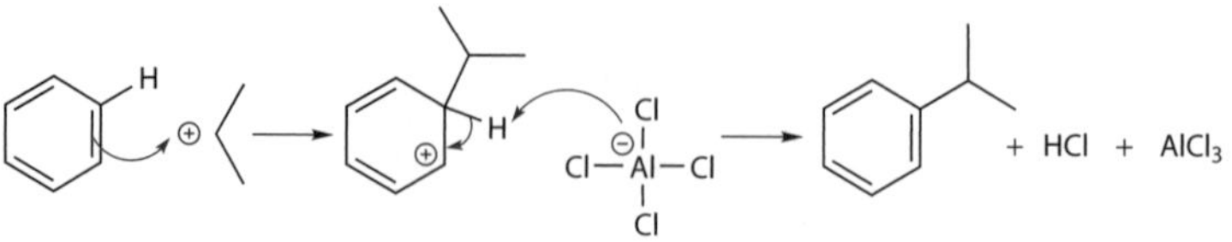

Friedel–Crafts acylation

- AlCl$_3$ is a strong Lewis acid and is required to generate the highly reactive acylium ion electrophile (Figure 7.29).

Figure 7.28
Friedel-Crafts acylation.

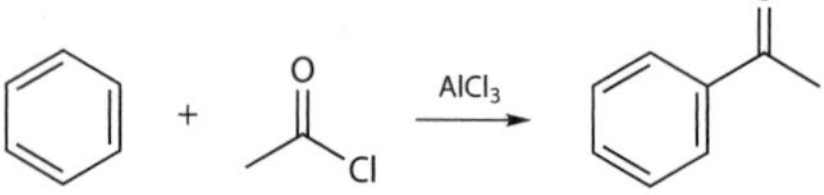

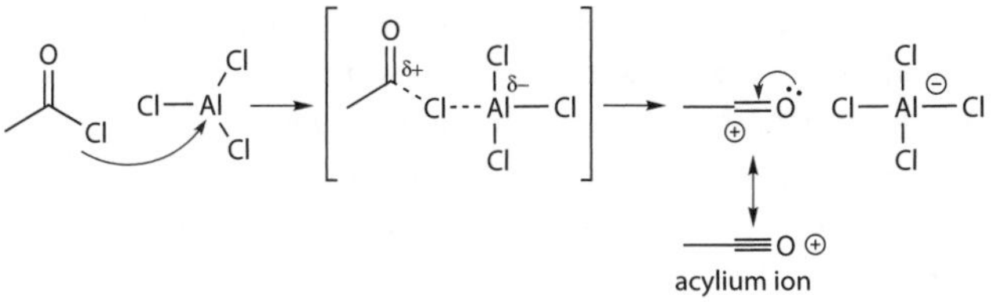

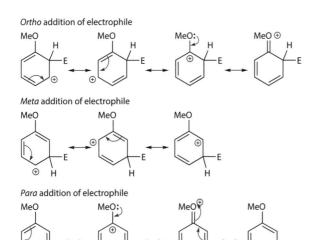

Figure 7.29
Formation of the electrophile.

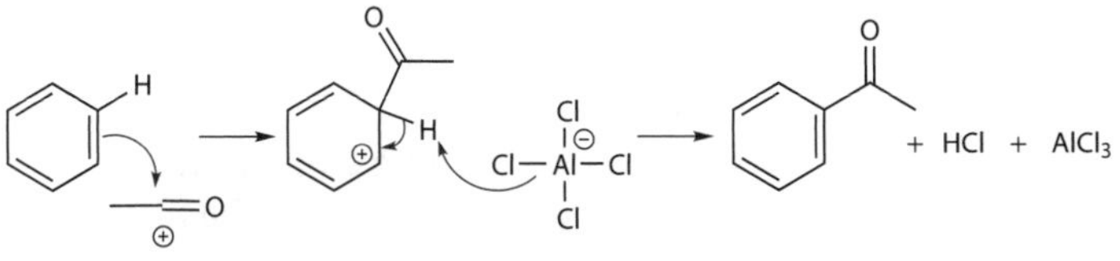

acylium ion

Figure 7.30
Reaction between benzene and the electrophile.

+ HCl + AlCl₃

- The acylium ion is stabilised by resonance which prevents the rearrangement observed in Friedel–Crafts alkylations.

Monosubstituted benzene derivatives

- If a substituent is present on the benzene ring it will affect both the rate of addition of an electrophile and where the electrophile adds.
- It is necessary to consider each of the three possible benzonium intermediates and evaluate the stability of each by comparing resonance stabilisation.
- In the case of the electron-donating methoxy group (Figure 7.31), an additional resonance structure is accessible when addition occurs at the *ortho* and *para* positions. This additional structure is formed from the interaction of a lone pair of electrons from the oxygen with the π system of the ring.

Figure 7.31
Electron-donating methoxy group.

Ortho addition of electrophile

Meta addition of electrophile

Para addition of electrophile

- For *meta* addition it is not possible to draw a canonical structure with the positive charge adjacent to the oxygen atom preventing the oxygen lone pairs from interacting with the π system.
- The additional resonance structure accessible from *ortho* and *para* addition provides additional stability to these benzonium intermediates and lowers the activation energy for their formation.

Tip

Electron-donating groups include:

–OH, –OR, –NH₂, –NHR, –NR₂

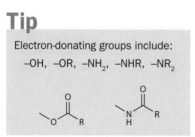

- Electron-donating substituents give both *ortho* and *para* products; there is no *meta* product formed.
- If the electron-donating group is bulky or if a bulky electrophile is used the reaction will favour formation of the *para* product. This is an example of steric hindrance driving selectivity in a reaction.
- Steric hindrance is only considered after electronic effects have been considered.

- In the case of the electron-withdrawing ester group (Figure 7.32) three different resonance structures are accessible for each site of addition. However, addition at the *ortho* and *para* positions leads to a resonance structure in which a positive charge is located adjacent to the ester, the carbon of which bears a δ+ due to the two adjacent electron-withdrawing oxygen atoms.

Figure 7.32 Electron-withdrawing ester group.

Ortho addition of electrophile

Meta addition of electrophile

Para addition of electrophile

This resonance structure is disfavoured and increases the activation energy for the formation of these addition products, making the benzonium originating from *meta* addition the most stable.

- Electron-withdrawing substituents give *meta* products.

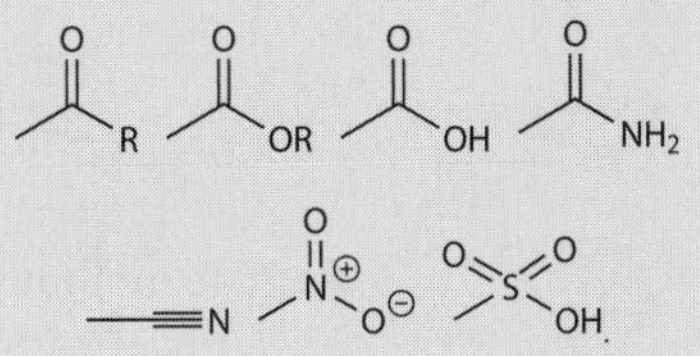

Halogen substituents

- Halogens, being more electronegative than carbon, withdraw electron density from an aromatic ring by induction and deactivate the ring as a result.
- The halogens have pairs of non-bonding electrons that can be pushed into the ring through resonance (Figure 7.33). In this way the halogens direct the addition of electrophiles to the *ortho* and *para* positions.

KeyPoints

- Electron-donating groups activate aromatic rings and are *ortho-*, *para*-directing.
- Electron-withdrawing groups deactivate aromatic rings and are *meta*-directing.
- Halogens deactivate aromatic rings but are *ortho-/para*-directing.

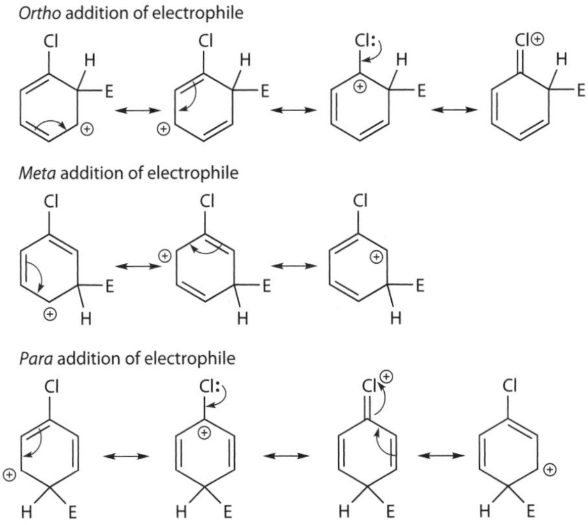

Figure 7.33
Halogens as an electron-donating group.

Ortho addition of electrophile

Meta addition of electrophile

Para addition of electrophile

Disubstituted benzene derivatives

- When there are two substituents on an aromatic ring, the directing effect of each substituent should be considered individually.
- The ketone is electron withdrawing and *meta* directing (blue arrows in Figure 7.34).

Figure 7.34
Electron-donating and withdrawing groups competing.

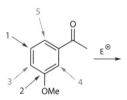

- The methoxy group is electron donating and *ortho/para* directing (grey arrows).
- Displacement of an existing group will not occur.
- Activating groups take priority over deactivating groups; there will be greater electron density at the *ortho* and *para* positions.
- Addition of an electrophile at position 4 would result in three substituents being next to each other; this is highly disfavoured on steric grounds.
- The electrophile will add at positions 3 and 5, giving a mixture of products.

Figure 7.35
Electron donating by induction and resonance competing.

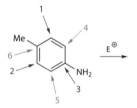

- The methyl group is electron donating by induction and *ortho/para* directing (blue arrows in Figure 7.35).
- The amino group is electron donating by resonance and *ortho/para* directing (grey arrows).
- Displacement of an existing group will not occur.
- Resonance effects are larger contributors than inductive effects.
- The electrophile will add at positions 4 and 5.
- Since the molecule is symmetric, addition at positions 4 and 5 will yield the same product.
- General rules for addition of electrophiles:
    - Electron-donating groups win over electron-withdrawing groups.
    - Resonance effect wins over inductive effect.
    - 1,2,3-Trisubstituted products are rarely formed due to steric crowding.
    - Bulky directing groups usually give more *para* substitution than *ortho* substitution.
    - The electrophile will replace a hydrogen; it will not replace an existing substituent.

Self-assessment

1. **Which of the following statements is incorrect?**
a. Aromatic compounds are planar.
b. Aromatic compounds have 4n π electrons.
c. Aromatic compounds are cyclic.
d. Aromatic compounds are less reactive than similarly substituted alkenes.

2. **In electrophilic aromatic substitution reactions, amino substituents are considered to be:**
a. *ortho*/*para* directing and activating
b. *ortho*/*para* directing and deactivating
c. *meta* directing and activating
d. *meta* directing and deactivating

3. **Which of the following compounds would be most reactive towards electrophilic substitution?**

a.

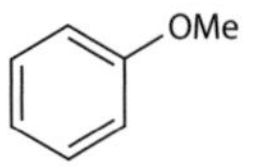

b.

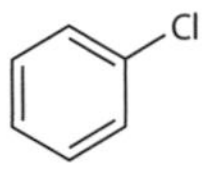

c.

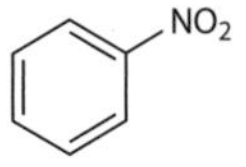

d.

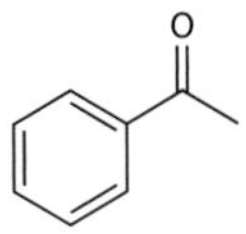

4. **Identify the major product of the following reaction:**

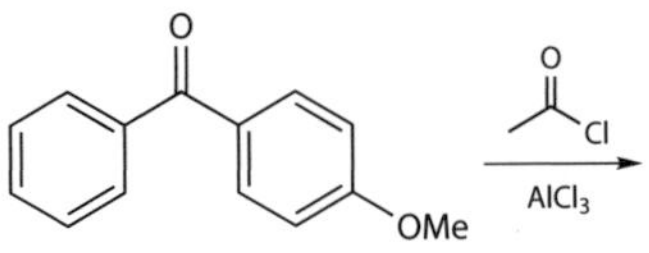

a.

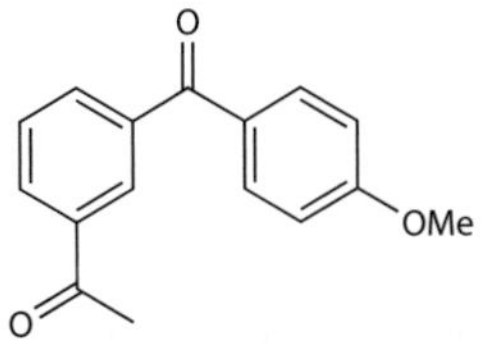

b.

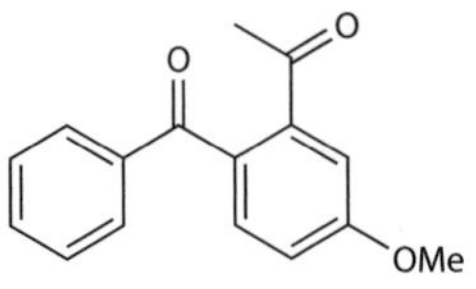

c.

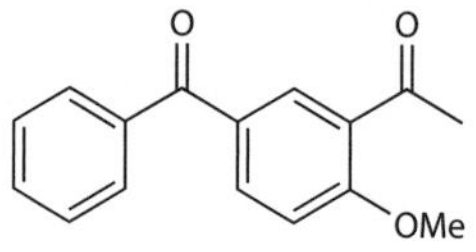

d.

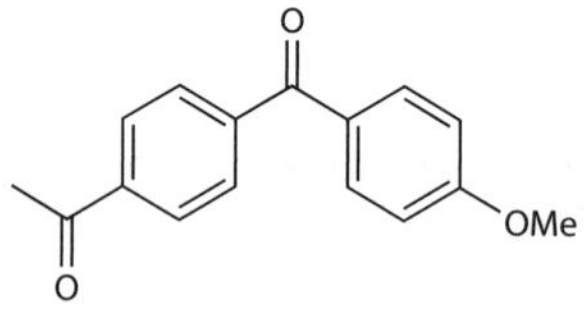

5.   **Identify the major product(s) of the following reaction:**

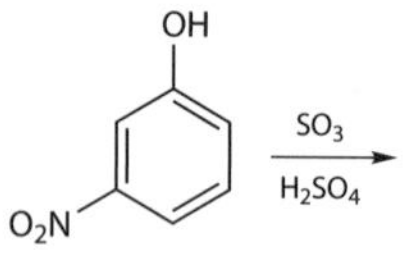

a.

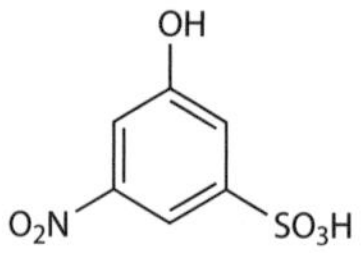

b.

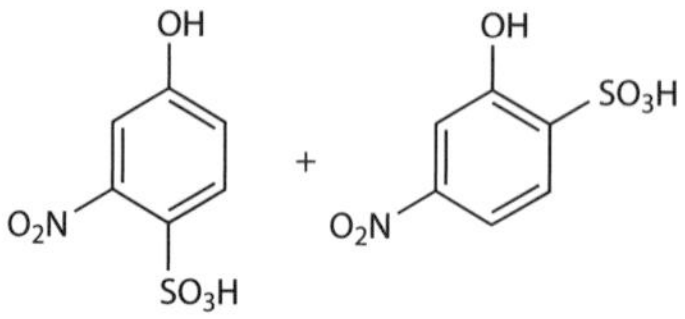

c.

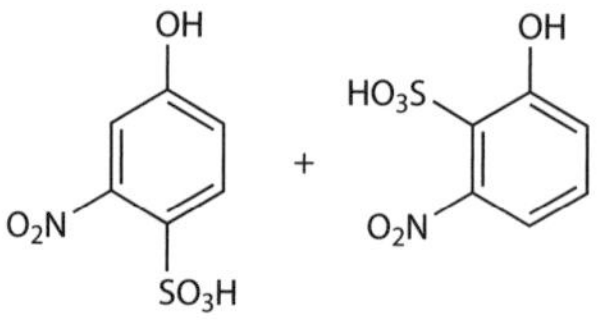

d.

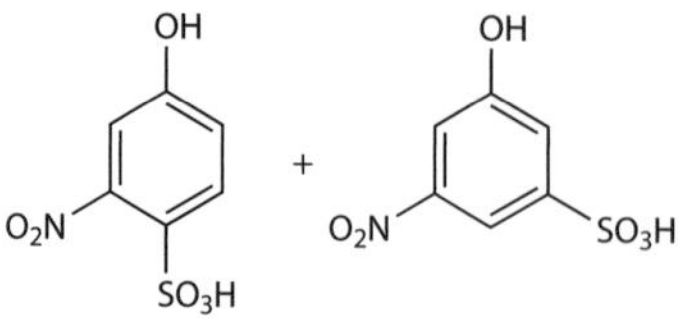

6.   **In Friedel–Crafts alkylations, which of the following statements is untrue?**
a.   Primary alkyl halides undergo rearrangement.
b.   A strong Lewis acid is required.

c.   The benzonium intermediate is positively charged.
d.   Benzene acts as the electrophilic component.

7.   **Identify the major product of the following reaction:**

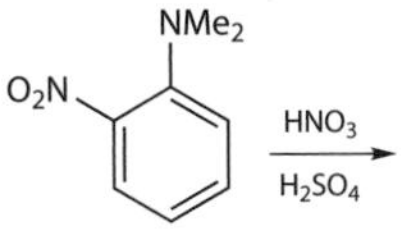

a.

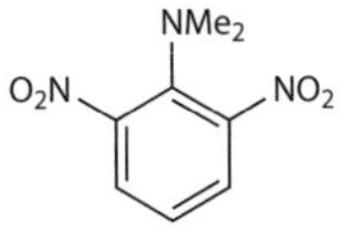

b.

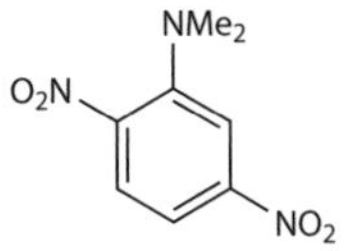

c.

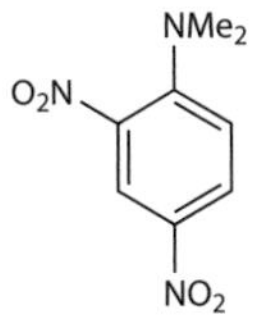

d.

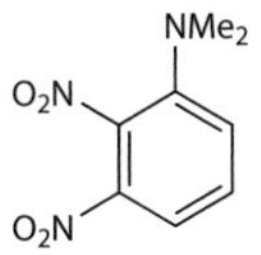

8.   **Which of the following compounds is aromatic?**

a.

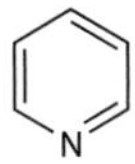

b.

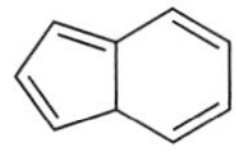

c.

d.

9. Which of the following is a key intermediate in the electrophilic aromatic substitution reaction mechanism?

a.

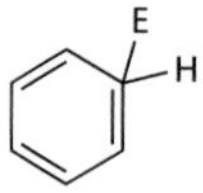

b.

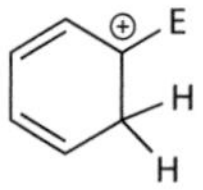

c.

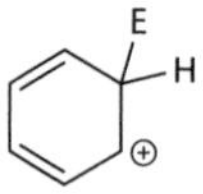

d.

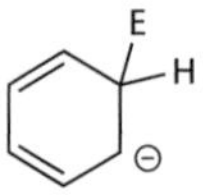

10. Identify the major product of the following reaction:

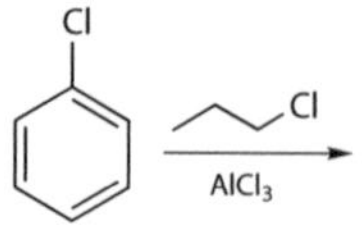

a.

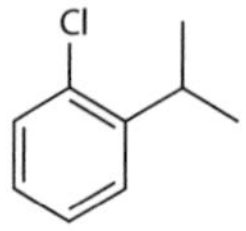

b.

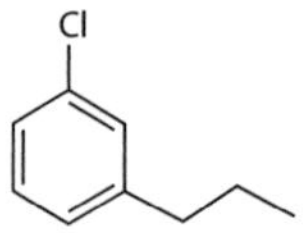

c.

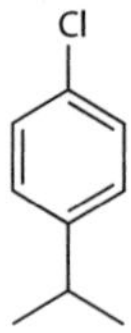

d.

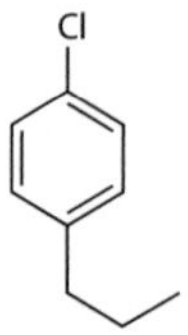

chapter 8

Chemistry of carbonyl compounds

Overview

After learning the material presented in this chapter you should:

- understand the reactivity of the carbonyl group and the diverse range of reactions that it undergoes
- understand the mechanisms for these reactions
- appreciate the acidity of the α-hydrogen atoms adjacent to the carbonyl group
- be able to discuss keto–enol tautomerism and the reactivity of enols and enolates
- be able to discuss conjugate addition as an alternative to direct addition at the carbonyl group
- appreciate how many of the reactions carried out in the laboratory have a biosynthetic equivalent.

Reactivity of the carbonyl group

- The carbonyl group has two distinct modes of reactivity. In the first, the carbonyl group behaves as an electrophile towards incoming nucleophiles. While in the second, the acidity of the α-hydrogen atoms adjacent to the carbonyl group allows the molecule to behave as a nucleophile (Figure 8.1).

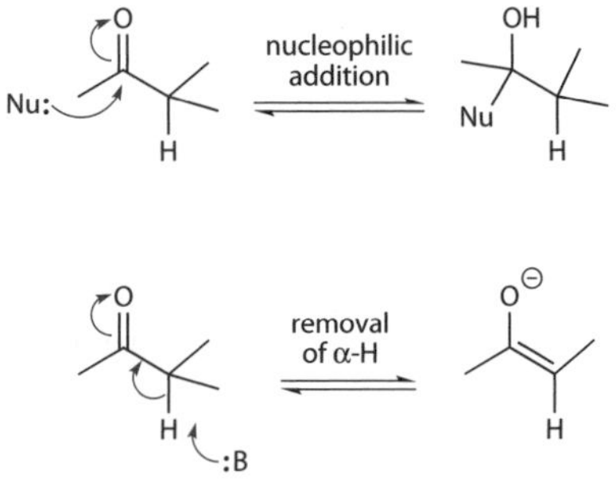

Figure 8.1
The electrophilic and nucleophilic modes of carbonyl group reactivity.

Nucleophilic addition

- The π bond of a carbonyl group is electron-rich with high electron density located above and below the plane of the carbon framework.
- The two electrons of the π bond are not equally shared by the carbon and oxygen atoms.

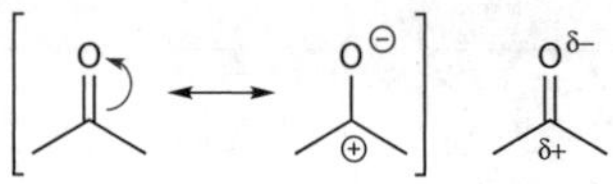

Figure 8.2
Resonance in carbonyl groups.

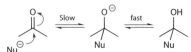

- The carbonyl group can adopt a charge-separated resonance hybrid where the negative charge is located on the more electronegative oxygen atom (Figure 8.2). Although this resonance hybrid is higher in energy than the original carbonyl group, it does give the carbon atom partial positive-charge character. As a result, carbonyl groups are electrophilic at carbon.
- When a nucleophile reacts with a carbonyl group the non-bonding electrons from the nucleophile attack the π bond, forming a new σ bond between the nucleophile and the carbon atom of the carbonyl group.
- The process of breaking the carbonyl bond results in a pair of electrons being localised on the oxygen atom. This intermediate anion is then quenched through protonation during acidic workup.
- Nucleophilic addition to a carbonyl group is an equilibrium process since the nucleophile is capable of acting as a leaving group (Figure 8.3).

Figure 8.3
Generalised mechanism for nucleophilic addition (basic conditions).

- The position of the equilibrium will be determined by the nucleophilicity of the nucleophile, the ability of the nucleophile to act as a leaving group and the structure of the carbonyl-containing molecule (Figure 8.4).

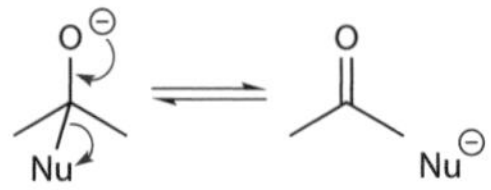

Figure 8.4
Reversibility of nucleophilic addition.

- Under acidic conditions the carbonyl group will be protonated through the donation of a pair of non-bonding electrons from oxygen.
- The resulting species is more electrophilic than the original carbonyl group and will react with weaker nucleophiles, including those without a formal negative charge (Figure 8.5).

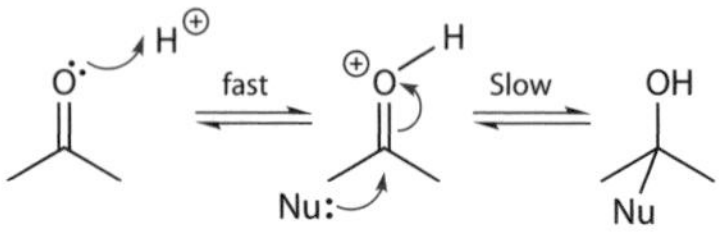

Figure 8.5
Generalised mechanism for nucleophilic addition (acidic conditions).

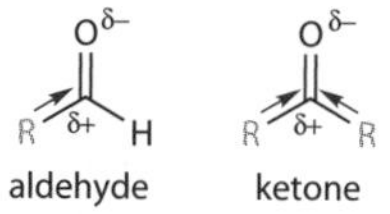

O$^{\delta-}$ O$^{\delta-}$

R $_{\delta+}$ H   R $_{\delta+}$ R

aldehyde   ketone

Figure 8.6 Electrophilic character of aldehydes versus ketones.

- Aldehydes are more reactive towards nucleophilic addition than ketones due to only having one electron-donating alkyl group attached to the carbonyl (Figure 8.6). The two alkyl groups on the ketone diminish the partial positive charge to a greater extent and reduce the electrophilicity of the ketone relative to the aldehyde.

Changes in hybridisation during nucleophilic addition

- Initially, both the carbon and the oxygen atoms of the carbonyl group are sp^2 hybridised. The non-bonding pair of electrons of the nucleophile attacks the π^* orbital of the carbonyl group at carbon, breaking the π bond and displacing the electrons from the π bond on to the oxygen atom.
- The attack of the nucleophile requires the hybridisation on carbon to change from sp^2 to sp^3 hybridisation to minimise the repulsion between the four pairs of bonding electrons, similarly the hybridisation on oxygen must change to one of sp^3 hybridisation to minimise non-bonding interactions.
- The protonation of the intermediate anion does not require a change in hybridisation at oxygen since one of the non-bonding pairs of electrons is being donated to the empty 1s orbital on hydrogen, forming a bonding pair of electrons (Figure 8.7).
- When the hybridisation of the carbonyl group changes to sp^3 hybridisation, the bond angle between the two groups attached to the carbonyl group is reduced from 120° to 109°.
- This reduction in bond angle will bring the two groups closer together. If the two groups are large this increase in steric crowding will be disfavoured and the position of the equilibrium will favour the starting sp^2 hybridised carbonyl over the sp^3 hybridised addition product.

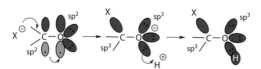

Figure 8.7
Changes in hybridisation.

Stereochemistry in nucleophilic addition

- If the carbonyl group bears two different substituents, the planar sp² hybridised framework will have enantiotopic faces since the nucleophile will be able to attack from either the top face (arrow a in Figure 8.8) or the bottom face (arrow b) to yield different enantiomeric products.
- The nucleophile has an equal probability of adding to the top face and the bottom face. Therefore, the two enantiomers will be formed in equal amounts, and a racemic mixture will result.

Figure 8.8
Nucleophilic attack at the sp² hybridised carbon.

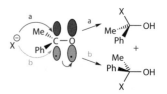

Examples of nucleophilic addition

Addition of cyanide

- The cyanide nucleophile will react with both aldehydes and ketones to form the corresponding cyanohydrin product.
- The reaction of aldehydes with cyanide will favour cyanohydrin formation due to both the increased reactivity of the aldehyde towards nucleophiles and the reduced steric hindrance encountered on forming the tetrahedral intermediate when compared with a ketone (Figure 8.9).

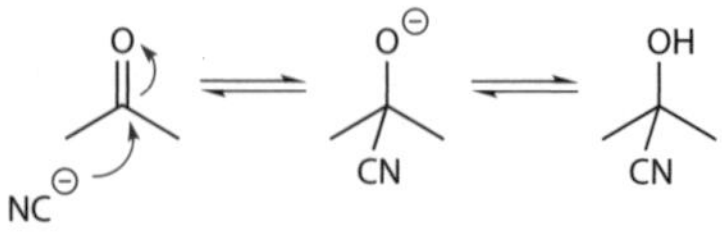

Figure 8.9
Nucleophilic addition of cyanide to a carbonyl.

Addition of Grignard reagents

- Grignard reagents are formed from the reaction of alkyl halides with magnesium in a coordinating, non-protic solvent (Figure 8.10).
- The high electropositivity of magnesium makes the carbon–magnesium bond highly polarised, giving significant nucleophilic character to the carbon atom of the bond.

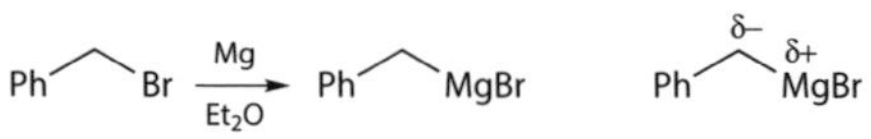

Figure 8.10
Grignard formation.

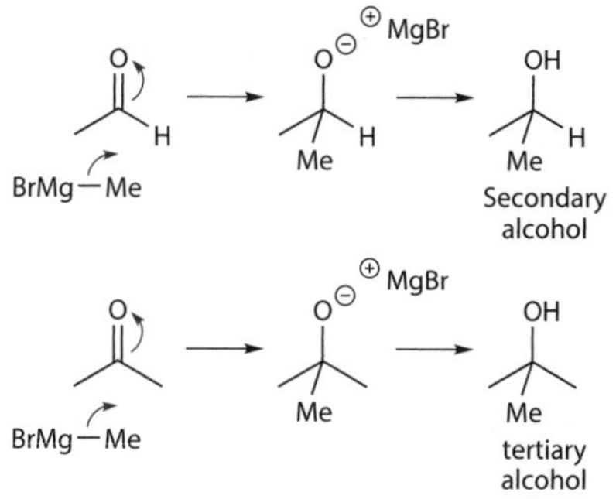

Figure 8.11
Nucleophilic addition of Grignard reagents to a carbonyl.

- Grignard reagents are potent nucleophiles.
- Grignard reagents react with both aldehydes and ketones, yielding secondary and tertiary alcohols respectively (Figure 8.11).
- The conversion of the magnesium salt to the product alcohol occurs on acidic workup in a second reaction step.
- The presence of acid or water in the first step would result in the decomposition of the Grignard reagent. The water is therefore only added to the reaction once the first step is complete (Figure 8.12).

Tip

Organolithium reagents are also a source of potent carbon nucleophile and react in the same way as Grignard reagents.

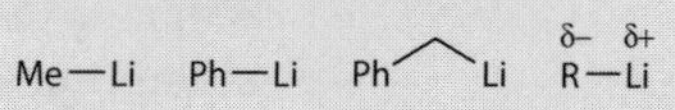

Me—Li   Ph—Li   Ph⌒Li $\overset{\delta-\ \ \delta+}{R—Li}$

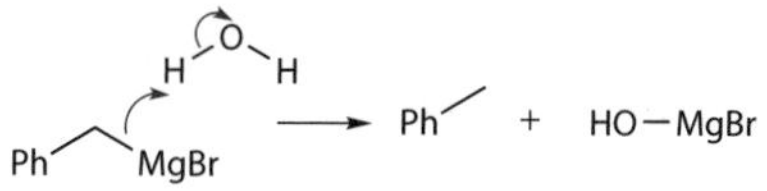

Figure 8.12
Hydrolysis of Grignard reagents.

Addition of hydride

- The small size and high charge density of the hydride anion (H⁻) means that it usually reacts as a base. Sodium hydride NaH is a powerful base.
- Sodium borohydride (NaBH$_4$) is a source of nucleophilic hydride. The delocalisation of electron density away from the hydrogen atoms in the borohydride anion (BH$_4^-$) allows for the transfer of hydride from the boron–hydrogen bond to an electrophile (Figure 8.13).

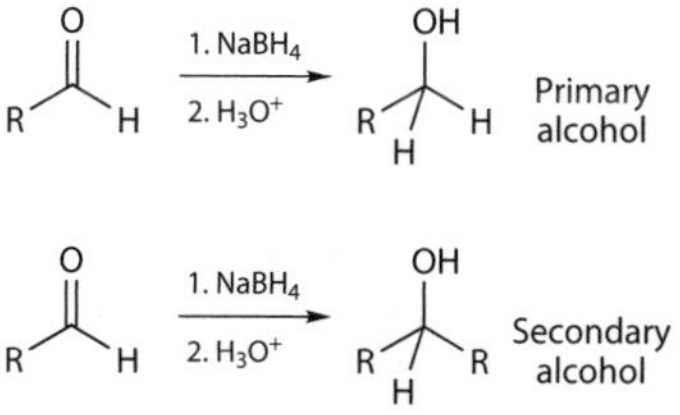

Figure 8.13
Nucleophilic addition of 'hydride' to a carbonyl.

Primary alcohol

Secondary alcohol

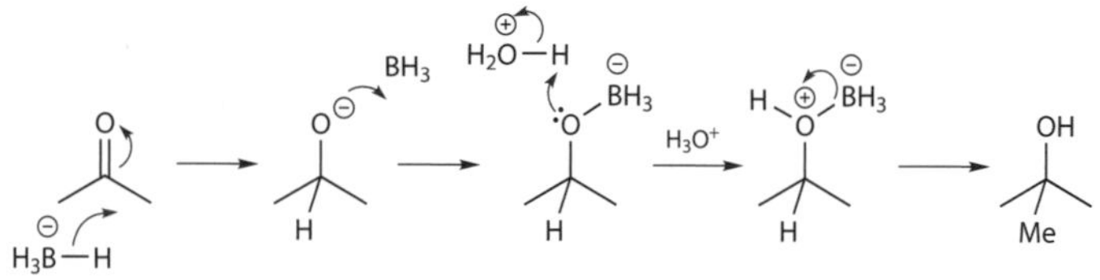

Figure 8.14
Mechanism for the reaction of NaBH$_4$ with a carbonyl.

- Sodium borohydride is a mild reducing agent, converting aldehydes and ketones to primary and secondary alcohols respectively.
- Sodium borohydride reduction will occur in the presence of water or alcohols, which will protonate the intermediate anion without destroying the reducing agent (Figure 8.14).
- The intermediate anion formed from the initial addition of hydride to the carbonyl group can stabilise the electron-deficient borane to generate a new boron species which can transfer a further three hydride to three additional carbonyl groups.
- Hydride reduction of carbonyl groups is an irreversible process since hydride is an exceptionally poor leaving group.
- Lithium aluminium hydride (LiAlH$_4$) is a more powerful reducing reagent due to the higher electropositivity of aluminium compared with boron; will also convert aldehydes and ketones to primary and secondary alcohols respectively.
- Lithium aluminium hydride reductions must be carried out under anhydrous reaction conditions as the reducing agent will rapidly react with water to form flammable hydrogen in a highly exothermic reaction.

Swern oxidation

- The reverse of hydride addition to carbonyl groups sees primary and secondary alcohols being converted to the corresponding aldehyde and ketone under oxidising conditions (Figure 8.15).

Figure 8.15
Oxidation of a primary alcohol to an aldehyde.

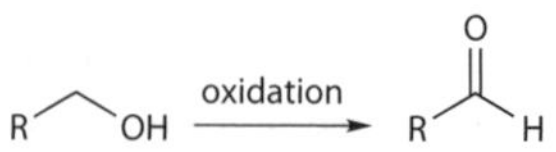

- Oxidising agents include potassium dichromate (K$_2$Cr$_2$O$_7$), potassium permanganate (KMnO$_4$), manganese dioxide (MnO$_2$) and dimethyl sulfoxide (DMSO).
- The use of DMSO in a Swern oxidation provides a useful route to aldehydes and ketones.
- Overoxidation of aldehyde products to the corresponding carboxylic acid does not occur under Swern conditions.
- The Swern oxidation consists of two steps. The first involves the reaction of DMSO with oxalyl chloride to form the chlorosulfonium ion (Figure 8.16).

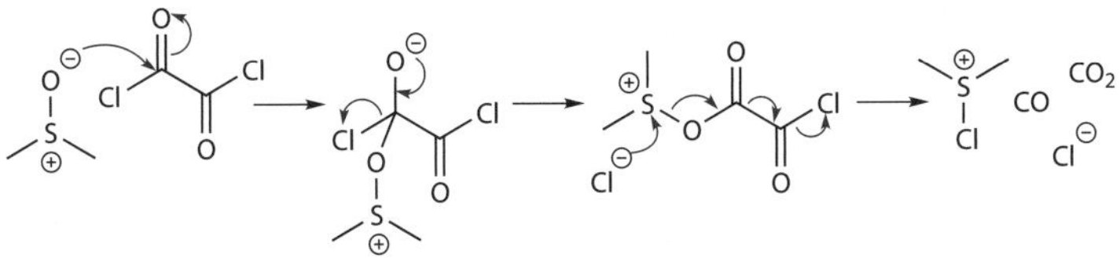

Figure 8.16
Formation of the electrophile.

- The chlorosulfonium ion then reacts with the carbonyl in the presence of triethylamine to abstract a hydrogen atom from the carbon adjacent to what was the hydroxyl group (Figure 8.17).

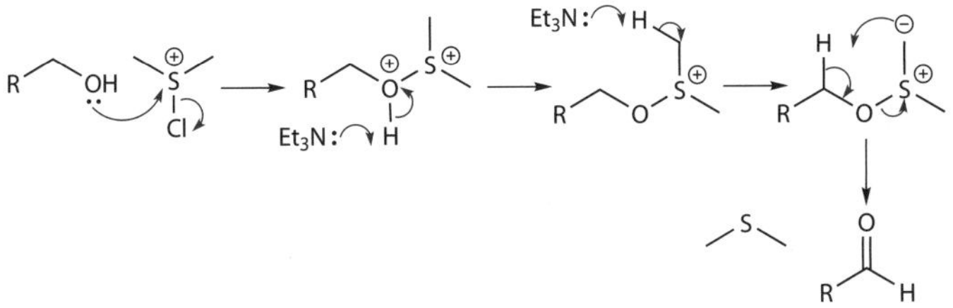

Figure 8.17
Reaction between an alcohol and the electrophile.

Addition of water – hydrate formation

- Aldehydes and ketones will react with water under acidic conditions following the initial protonation of the carbonyl group (Figure 8.18).
- The equilibria for hydrate formation lie heavily in favour of the starting carbonyl due to the increased steric hindrance of the sp^3 hybridised carbon of the product and the good leaving group characteristics of the water molecule under acidic reaction conditions.
- For aldehydes the equilibria will be more favourable towards hydrate formation than for ketones.

Tips

- Structural features can stabilise hydrates.
- In the case of chloral hydrate, two hydrogen bonds between the hydroxyls and nearby chlorine atoms stabilise the hydrate.

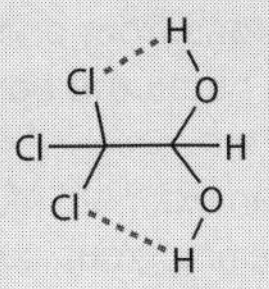

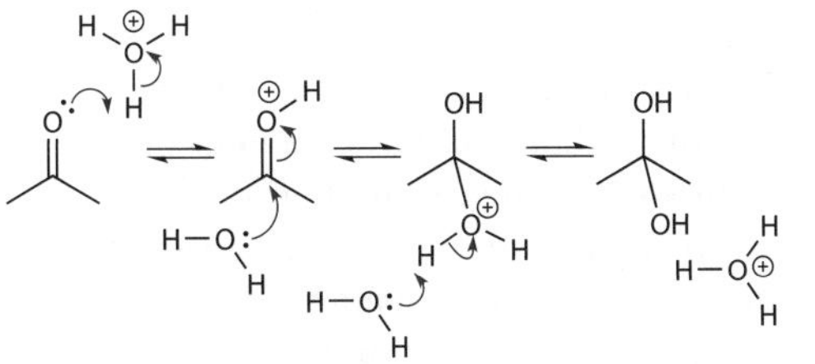

Figure 8.18
Nucleophilic addition of water to a carbonyl.

Figure 8.19
Nucleophilic addition of
alcohol to a carbonyl.

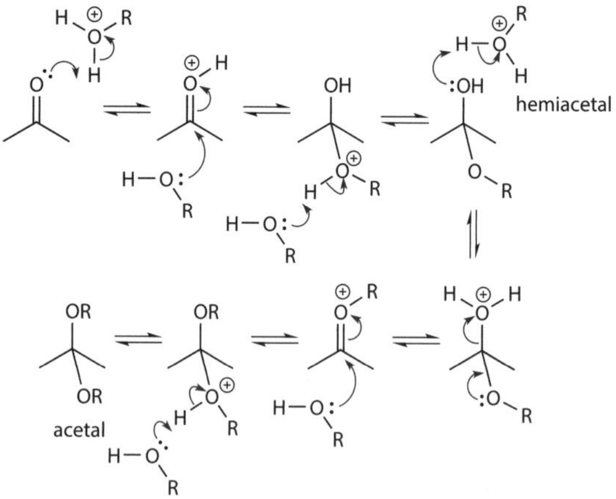

Addition of alcohols – hemiacetal and acetal formation

- Aldehydes and ketones will react with alcohols under acidic conditions to form a hemiacetal; this hemiacetal will react with a further molecule of alcohol to form an acetal (Figure 8.19).
- The electrophile used to form the hemiacetal is a protonated carbonyl and the electrophile used for the conversion of the hemiacetal to the acetal is an analogous species with a positive charge on oxygen.
- Acetal formation reactions can be driven to completion through the removal of water from the reaction; this pulls the individual equilibria to the right, favouring product formation.
- Hemiacetals are unstable with the equilibria favouring the carbonyl group. The exception are sugars, which are stable hemiacetals (Figure 8.20).

Figure 8.20
Glucose – a stable
hemiacetal.

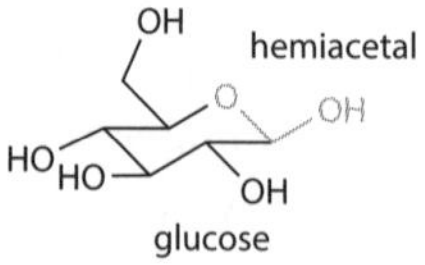

glucose

- Diols will react with carbonyl compounds to form cyclic acetals.
- The first hydroxyl of the diol will react with the carbonyl to form a hemiacetal (Figure 8.21). The second hydroxyl will then react in an intramolecular reaction since this hydroxyl is being held in close proximity to the electrophilic site.

KeyPoints

- Notice the position of the two oxygen atoms of the hemiacetal (Figure 8.19).
- When atoms bearing lone pairs of electrons are located in a 1,3-relationship, the lone pair of electrons from one atom can push to break the bond to the other atom.
- This is an example of neighbouring group participation.

Tips

- Intramolecular reactions have their reacting sites within the same molecule.
- Intermolecular reactions occur between reactive sites on two different molecules.

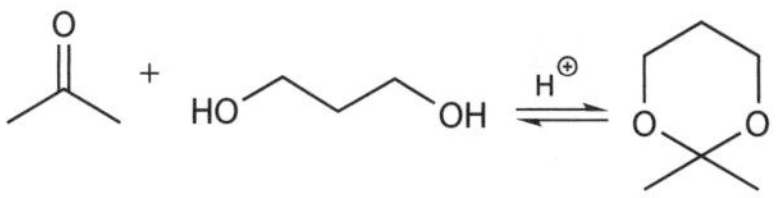

Figure 8.21
Nucleophilic addition of diols to a carbonyl.

- Acetals are not stable under acidic aqueous conditions, but they are stable under basic aqueous conditions.
- Drug molecules containing a hydroxyl group are converted to the more water soluble glycoside by the liver to aid with their excretion from the body.
- The glucose hemiacetal is converted to a glycoside acetal (Figure 8.22).

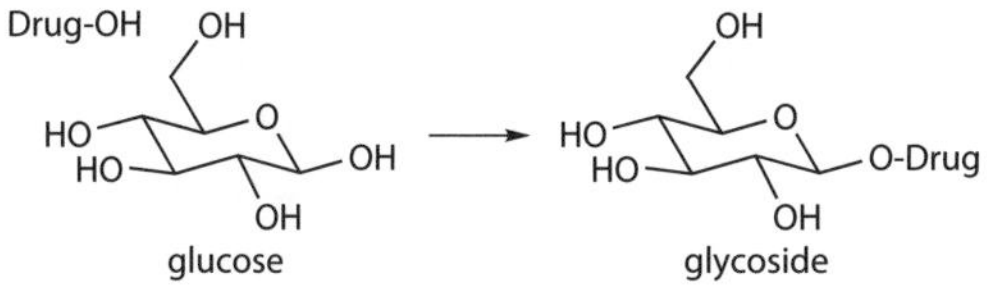

Figure 8.22
Glycoside formation in the liver.

glucose

glycoside

Addition of amines – imine and enamine formation

- Aldehydes and ketones will react with primary amines under both neutral and acidic conditions to form an imine (Figure 8.23).
- Imine formation is directly analogous to acetal formation, with the lone pair of electrons on nitrogen facilitating the elimination of the leaving group.
- Secondary amines reacting with aldehydes or ketones will not form an imine since there are no hydrogen atoms available on the nitrogen atom for deprotonation of the iminium in the final step.

KeyPoints

- Test for the presence of aldehydes and ketones.
- Treating the carbonyl-containing compound with 2,4-dinitrophenylhydrazine results in a yellow, orange or red precipitate being formed.

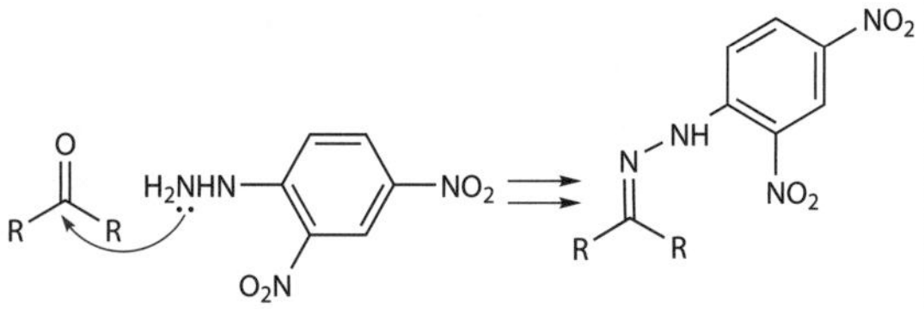

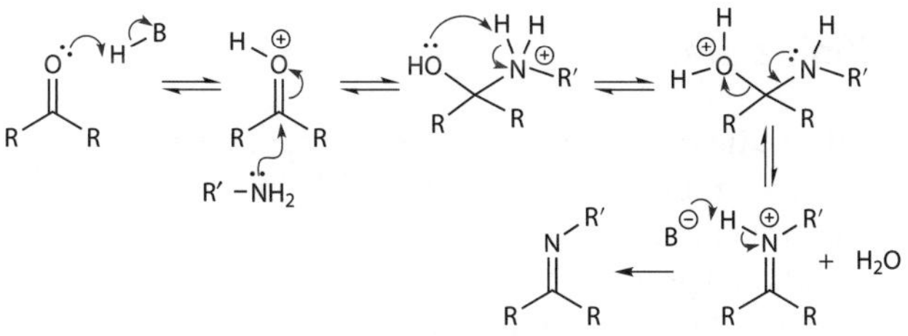

Figure 8.23
Nucleophilic addition of primary amines to a carbonyl.

Figure 8.24
Nucleophilic addition of secondary amines to a carbonyl.

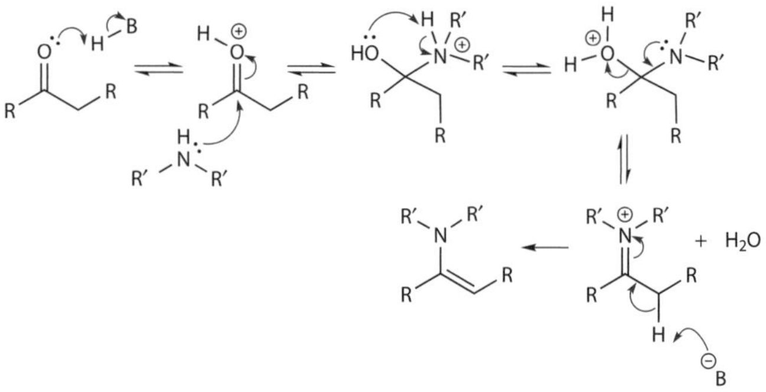

- The intermediate iminium will instead form an enamine through deprotonation on an adjacent carbon atom.
- The electrons from the carbon–hydrogen bond form a new carbon–carbon π bond with the carbon–nitrogen π bond breaking to quench the positive charge on nitrogen (Figure 8.24).

KeyPoints

- Typically the iminium formation will be carried out in the presence of the reducing agent in a 'one-pot' reaction.
- It is essential that the reducing agent that is used will selectively reduce the iminium and not unreacted carbonyl.
- Sodium cyanoborohydride (NaBH$_3$CN) is a selective reducing agent for this reaction.

Reductive amination

- In the same way that carbonyl groups can be reduced to the corresponding alcohol, iminium salts can be reduced in the presence of a hydride source to the corresponding amine.
- The process of converting a carbonyl group to an amine is termed reductive amination and is a powerful tool in organic synthesis (Figure 8.25).

Figure 8.25
Reductive amination of a carbonyl group.

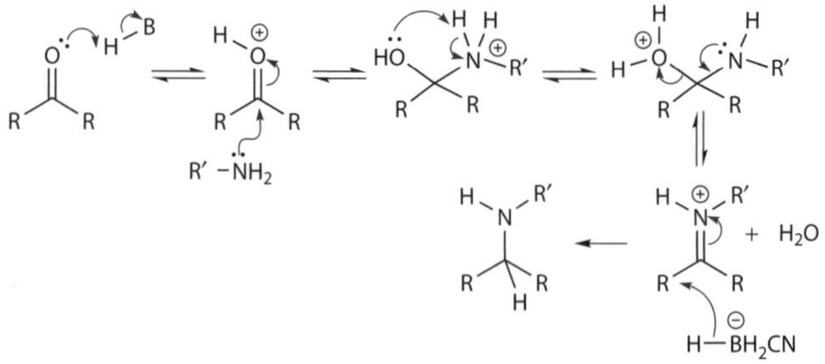

Addition–elimination

- When a nucleophile reacts with a carbonyl group, the non-bonding electrons from the nucleophile attack the π bond, forming a new σ bond between the nucleophile and the carbon atom of the carbonyl group.
- The process of breaking the carbonyl bond results in a pair of electrons being localised on the oxygen atom.

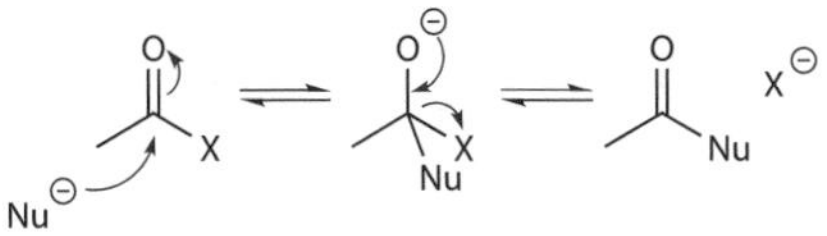

Figure 8.26
Generalised mechanism for addition–
elimination (basic conditions).

- If there is a suitable leaving group on the carbon, the tetrahedral intermediate will collapse, with a non-bonding pair of electrons from the oxygen reforming the carbonyl π bond and displacing the electrons from the σ bond on to the leaving group (Figure 8.26).
- This reaction pathway is distinct to nucleophilic addition at the carbonyl. In nucleophilic addition there is no leaving group available for the elimination step.
- Addition–elimination reactions can proceed under acidic, basic or neutral reaction conditions.
- Although you can draw curly arrows for this process such that the reaction appears to proceed through an S_N2 mechanism, this is incorrect. The empty π^* acceptor orbital on the carbonyl is lower in energy than the σ^* acceptor orbital of the carbon-leaving group bond.
- The approaching nucleophile would also need to attack the carbon of the electrophile in the plane of the carbonyl group for an S_N2 mechanism. This approach trajectory would be highly congested (Figure 8.27).

Tip

No S_N2 at sp^2 – ever!

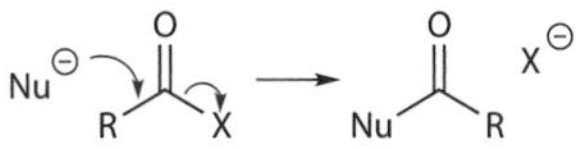

Figure 8.27
Disallowed S_N2 attack at the sp^2 carbon.

- Addition–elimination reactions are reversible due to the availability of two leaving groups in the tetrahedral intermediate. Either the leaving group initially attached to the carbonyl or the incoming nucleophile can be eliminated (Figure 8.28).
- Factors affecting product formation:
  1. nucleophilicity of the incoming nucleophile
  2. reactivity of the substrate
  3. relative leaving group ability of the nucleophile and the group X
  4. reactivity of the product relative to the starting material.

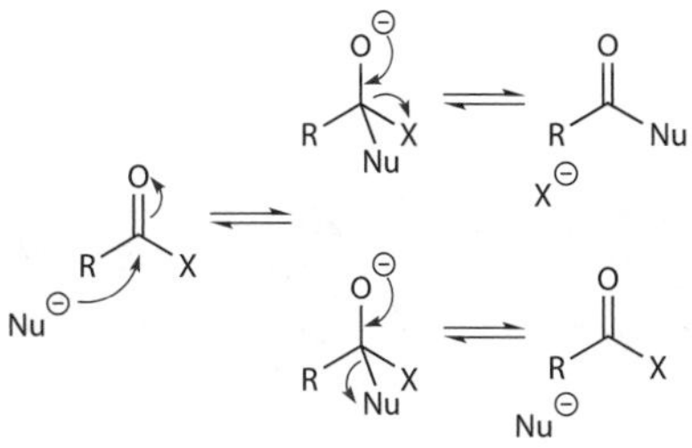

Figure 8.28
Reversibility of addition–elimination reactions.

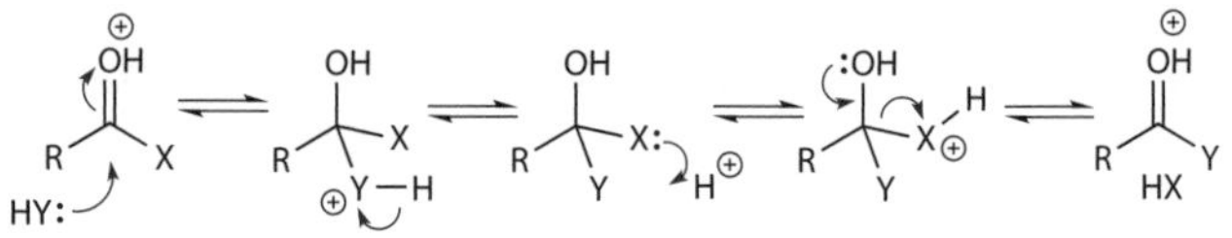

Figure 8.29
Generalised mechanism for addition–elimination (acidic conditions).

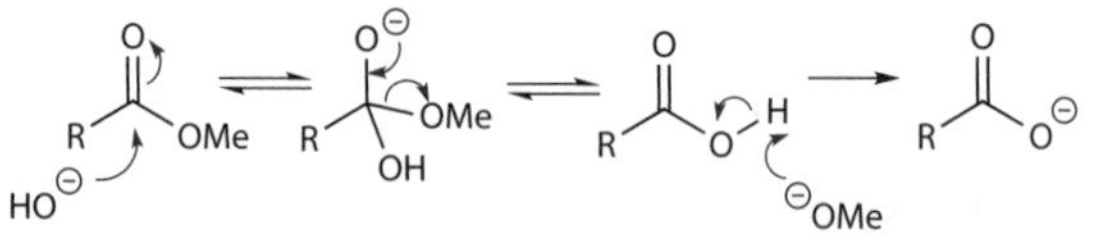

Figure 8.30
Irreversible steps drive equilibria towards product formation.

- Under acidic conditions the protonation of the carbonyl group to generate a more powerful electrophile will promote reactions with poor nucleophiles (Figure 8.29).
- The reversibility of addition–elimination reactions can be shifted through either the removal of the product X/HX to drive the equilibria to favour product formation or through the inclusion of an irreversible reaction step.
- Under basic reaction conditions, ester hydrolysis proceeds through a series of reversible steps to generate a carboxylic acid product. Under the reaction conditions this is deprotonated to the carboxylate and prevents the ester from being reformed (Figure 8.30).

Changes in hybridisation during addition–elimination

- Initially, both the carbon and oxygen atoms of the carbonyl group are sp² hybridised. The non-bonding pair of electrons of the nucleophile attacks the π* orbital of the carbonyl group at carbon, breaking the π bond and displacing the electrons from the π bond on to the oxygen atom.
- The attack of the nucleophile requires the hybridisation on carbon to change from sp² to sp³ hybridisation to minimise the repulsion between the four pairs of bonding electrons. Similarly, the hybridisation on oxygen must change to one of sp³ hybridisation to minimise non-bonding interactions.
- When the non-bonding pair of electrons reform the π bond to eliminate the leaving group both the oxygen and the carbon must revert back to sp² hybridisation.
- The leaving group takes the pair of electrons that were previously bonding the group to the carbon.
- Overall there is no change in hybridisation at either the carbon or the oxygen (Figure 8.31).

Figure 8.31
Changes in hybridisation.

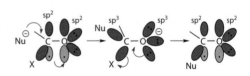

Examples of addition–elimination

■ Addition–elimination reactions will take place with esters, amides, acyl halides and anhydrides. The reactivity of these substrates is determined by the basicity of their respective leaving groups.

■ The more basic the leaving group, the less stable the group and the less likely it will be to leave the tetrahedral intermediate (Figure 8.32).

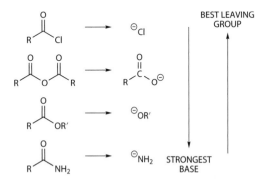

BEST LEAVING GROUP

STRONGEST BASE

Figure 8.32
Leaving group stability for carboxyl derivatives.

■ Acyl halides and anhydrides will react with nucleophiles without acid or base catalysis.

■ Esters and amides are significantly less reactive towards nucleophilic attack and require either acid or base catalysis to enhance either the nucleophilicity of the nucleophile or the electrophilicity of the substrate.

■ Esters and amides will typically require forcing reaction conditions for a reaction to occur.

■ The reduced reactivity of esters and amides towards nucleophilic attack is due to the delocalisation of electron density in these groups (Figure 8.33).

■ In the case of esters, the high electronegativity of oxygen makes the non-bonding pair of electrons on oxygen less available for resonance than the corresponding electrons on the amide nitrogen. The ester has more 'carbonyl' character than the amide.

■ The reduced resonance stabilisation of the ester compared with the amide gives the ester increased electrophilic character, making it more reactive towards nucleophiles.

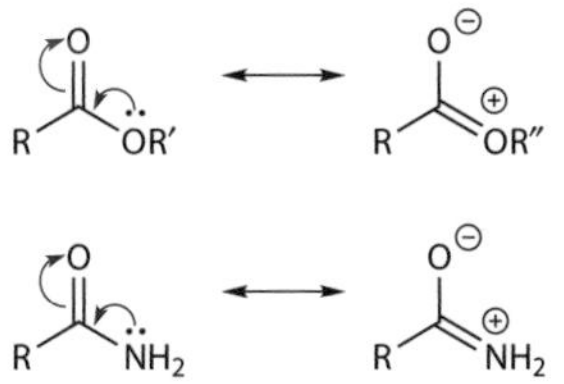

Figure 8.33
Resonance in esters and amides.

Hydrolysis, esterification and amide formation

- Acyl halides, anhydrides and esters will react with water, alcohols and amines to yield the corresponding carboxylic acid, ester and amide (Figure 8.34 and Table 8.1).
- Amides are much less reactive but will undergo hydrolysis to the carboxylic acid upon heating.
- For many reaction mechanisms it is possible to combine a number of individual steps into a single step through the use of multiple curly arrows (Figures 8.35 and 8.36). This approach will more accurately represent the concerted nature of many classes of reaction.

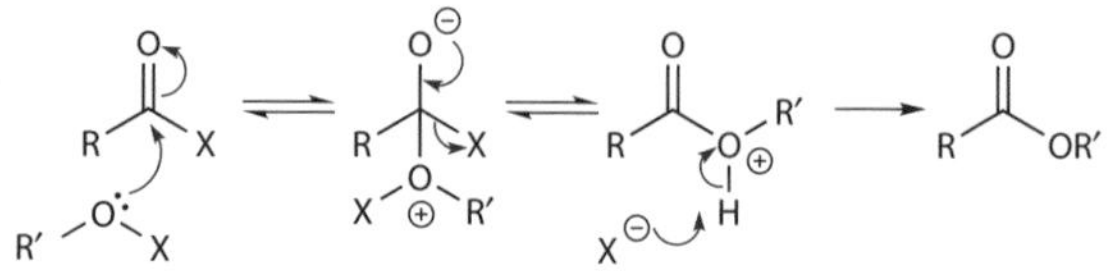

Figure 8.34
Mechanism for the addition of an alcohol to either an acyl halide or anhydride.

Table 8.1 Reaction of carboxyl derivatives with different nucleophiles

	H—OH	H—OR″	H—NR″₂	
R—C(=O)—Cl	R—C(=O)—OH	R—C(=O)—OR″	R—C(=O)—NR″₂	
R—C(=O)—O—C(=O)—R	R—C(=O)—OH	R—C(=O)—OR″	R—C(=O)—NR″₂	
R—C(=O)—OR	R—C(=O)—OH	R—C(=O)—OR″	R—C(=O)—NR″₂	Acidic conditions
	R—C(=O)—O⁻	R—C(=O)—OR″	R—C(=O)—NR″₂	Basic conditions

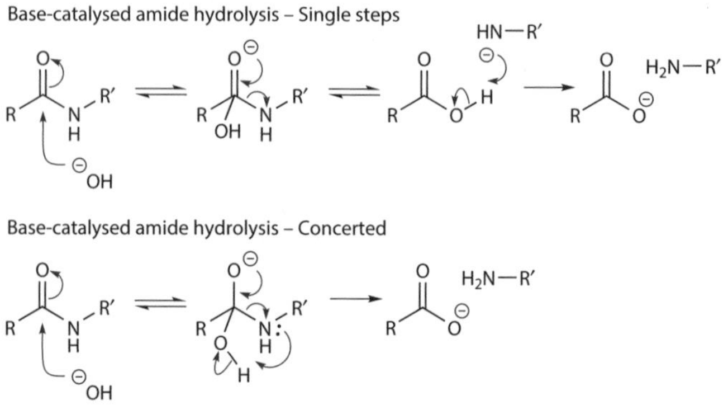

Figure 8.35
Comparison of drawing reaction mechanisms (base catalysis).

Base-catalysed amide hydrolysis – Single steps

Base-catalysed amide hydrolysis – Concerted

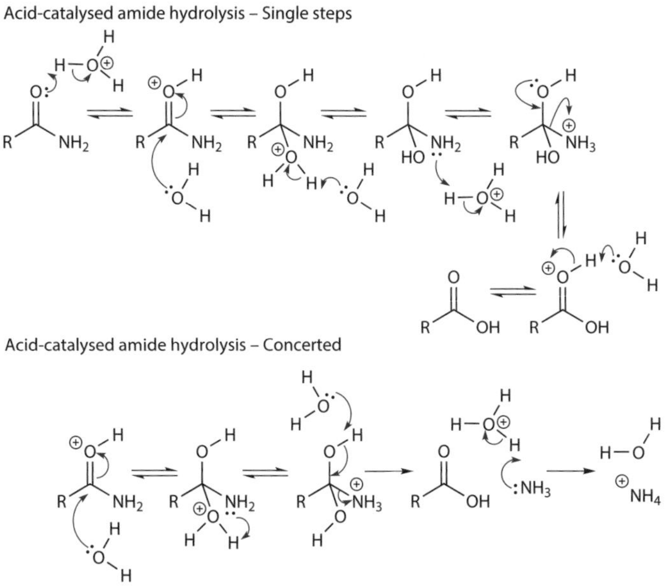

Acid-catalysed amide hydrolysis – Single steps

Figure 8.36
Comparison of drawing reaction mechanisms (acid catalysis).

Acid-catalysed amide hydrolysis – Concerted

- Both the single-step and concerted method of drawing a reaction mechanism are correct.
- Although amides require heating and either strongly acidic or basic conditions for hydrolysis in the laboratory, biology uses serine proteases to facilitate the same conversion under physiological conditions.
- The active site of these enzymes contains a conserved three-amino-acid catalytic triad which operates as a charge relay system to hydrolyse peptides selectively (Figure 8.37).

Tips

- When combining mechanistic steps as in Figures 8.34–8.36 there is a danger that the subtleties of a reaction may be lost.
- It is also much easier to lose track of electrons and charges when there are more arrows associated with a molecule.

Transesterification

- In the presence of alcohols, under both acidic and basic conditions, esters undergo transesterification (Figure 8.38), whereby the alkoxy group of the ester is exchanged with the alcohol to give a new ester (Figure 8.39).

Fischer esterification

- Fischer esterification (Figure 8.40) can be considered to be the reverse of ester hydrolysis. Under acidic conditions carboxylic acids react with alcohols to form the corresponding ester (Figure 8.41).

KeyPoints

- Esterification of carboxylic acids will only occur under acidic conditions.
- Under basic conditions the nucleophile will behave as a base to deprotonate the carboxylic acid.

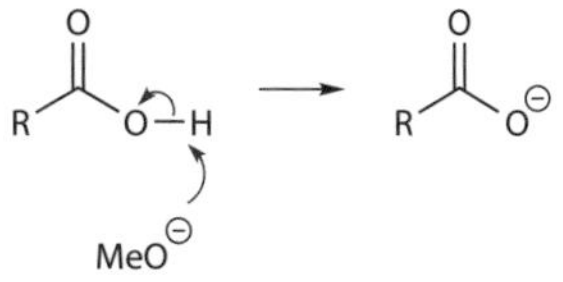

Figure 8.37
Mechanism of hydrolysis
within a serine protease.

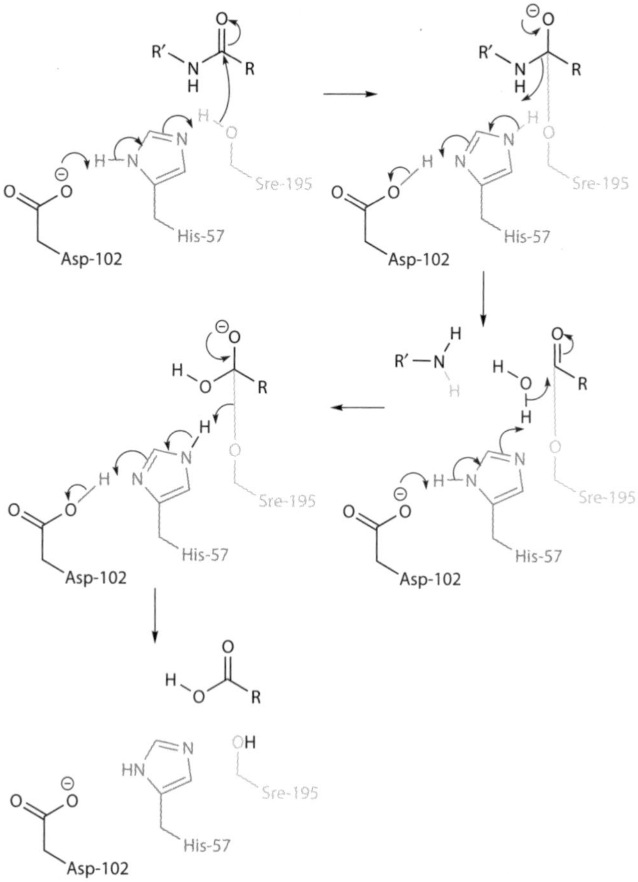

Figure 8.38
Transesterification.

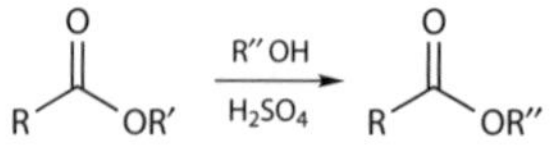

Figure 8.39
Addition–elimination of an
alcohol to an ester.

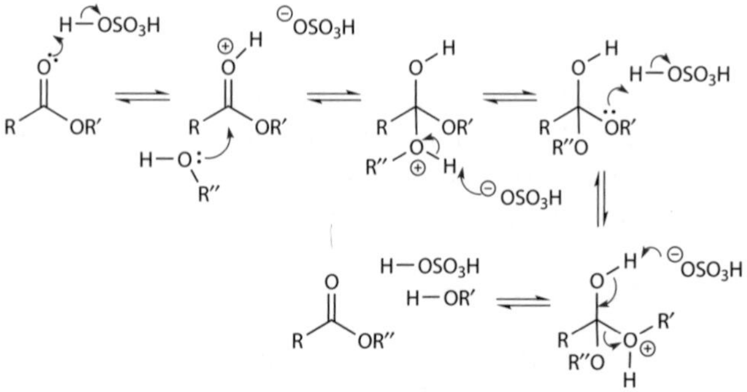

Figure 8.40
Fischer esterification.

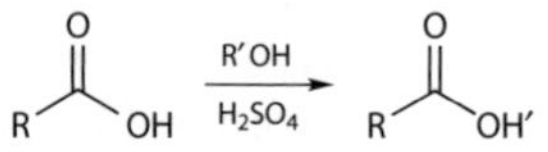

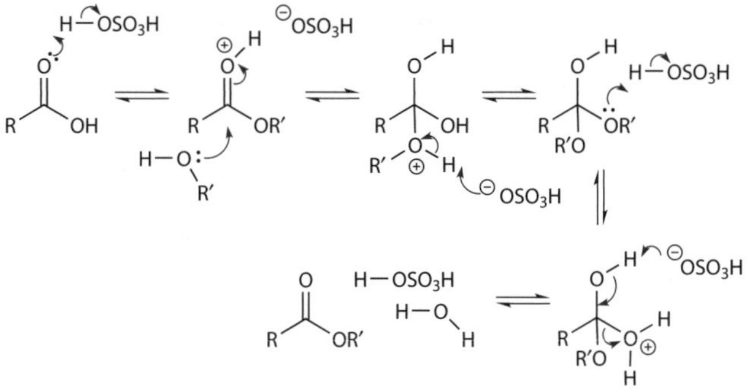

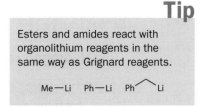

Figure 8.41
Addition–elimination of an
alcohol to a carboxylic acid.

Reaction with Grignard reagents

- Esters will react with Grignard reagents in the same way as aldehydes and ketones.
- Since esters have an available leaving group, the initial addition product will collapse to generate a ketone.
- The ketone, being more reactive than the ester towards nucleophiles, will then react with a second molecule of Grignard reagent to give a tertiary alcohol product (Figure 8.42).

Tip

Esters and amides react with organolithium reagents in the same way as Grignard reagents.

Me—Li   Ph—Li   Ph⌒Li

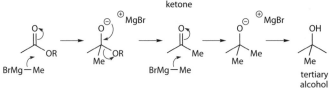

Figure 8.42
Addition–elimination of
Grignard reagents to an
ester.

Hydride reduction

- Esters and amides are not readily reduced by sodium borohydride but can be reduced with the more powerful lithium aluminium hydride to the corresponding primary alcohol.
- The reaction proceeds through the formation of an intermediate aldehyde, which, under the highly reducing reaction conditions, is immediately converted to the primary alcohol (Figure 8.43).

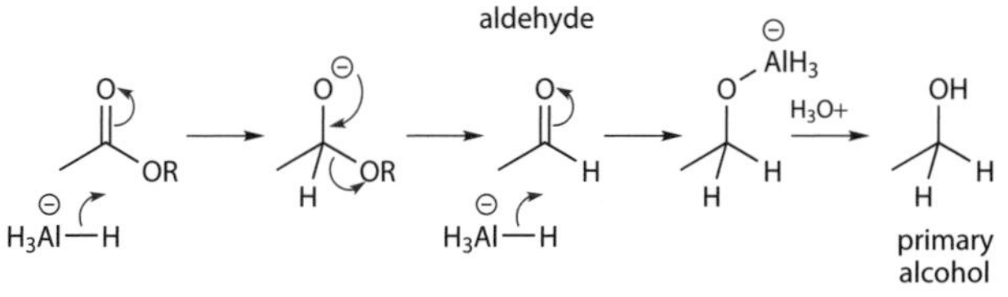

Figure 8.43
Addition–elimination of
'hydride' to an ester.

KeyPoints

- Test for the presence of esters.
- Treating the ester with hydroxylamine followed by iron(III) chloride results in a deep red solution being formed.
- The hydroxamate chelates iron to form an octahedral complex.

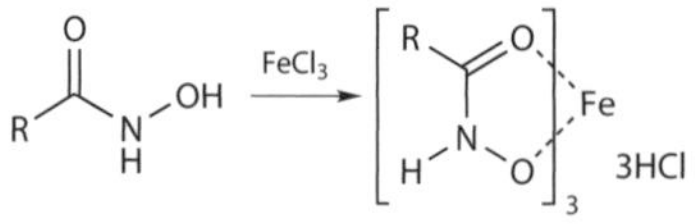

Reaction with hydroxylamine

- Hydroxylamine has non-bonding pairs of electrons on both nitrogen and oxygen. The electrons on oxygen will be more tightly bound due to the higher electronegativity of oxygen, making the nitrogen the more nucleophilic.
- Esters react with hydroxylamine to form the corresponding hydroxamate (Figure 8.44).

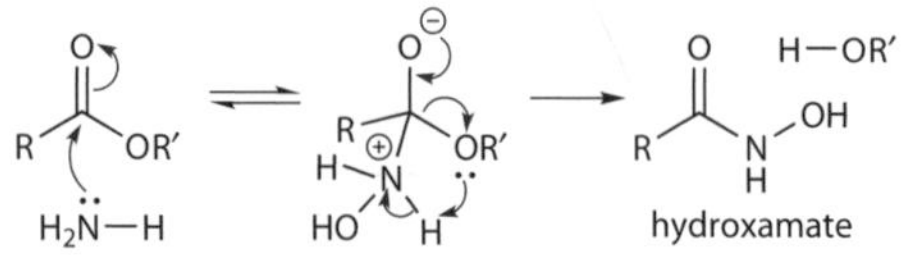

Figure 8.44
Addition–elimination of hydroxylamine to an ester.

α-Hydrogen reactivity

- The hydrogen atoms on the carbon adjacent to a carbonyl group, the α-hydrogen atoms, are significantly more acidic than the hydrogen atoms of a simple alkane.
- The acidity of the α-hydrogen is due to the ability to delocalise the negative charge of the conjugate base on to the carbonyl to give a resonance-stabilised enolate anion (Figure 8.45).

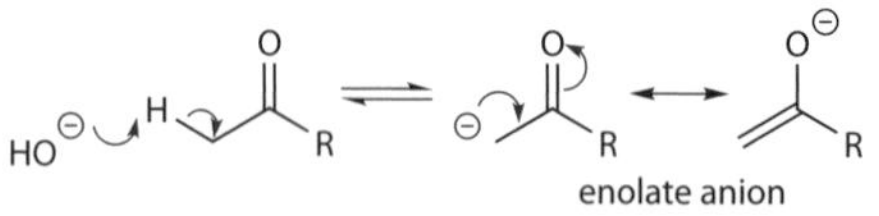

Figure 8.45
Acidity of α-hydrogen atoms.

- The most stable resonance hybrid will be the hybrid with the negative charge localised on the more electronegative oxygen atom. This is the hybrid to use when thinking about enolate reactivity (Figure 8.46).
- Under acidic conditions the carbonyl group can be converted to the analogous enol.
- Initial protonation of the oxygen of the carbonyl group makes this group a better electrophile for the electrons from the α-hydrogen–carbon bond.
- Upon deprotonation of the α-hydrogen an enol is formed.

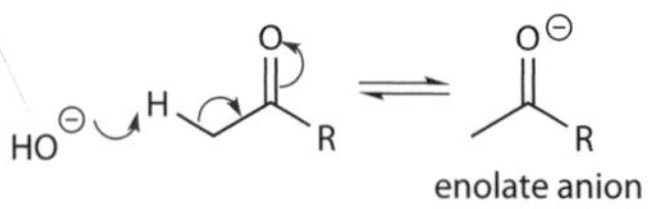

Figure 8.46
Enolate formation.

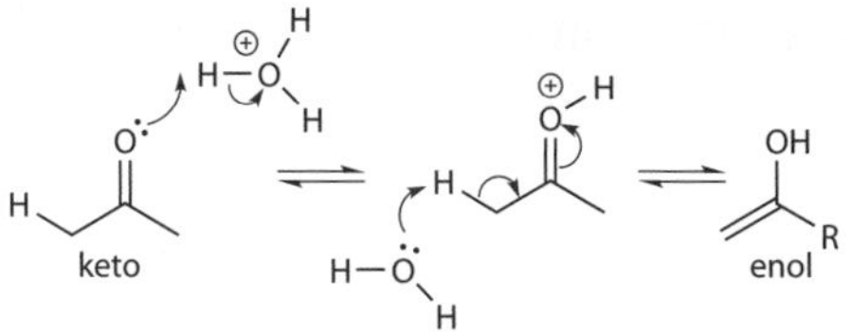

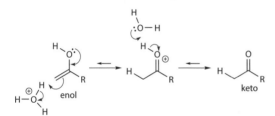

Figure 8.47
Keto–enol tautomerism.

KeyPoints

- Tautomers are species that exist in equilibrium and differ in the position of a proton.
- Tautomers are interconverted by chemical reactions.
- Tautomers are not resonance hybrids.

- The net result is the relocation of a hydrogen atom from carbon in the initial molecule to the oxygen of the enol. The molecule is said to exhibit keto–enol tautomerism (Figure 8.47).
- The combination of a carbon–oxygen double bond and a carbon–hydrogen single bond is lower in energy than a carbon–carbon double bond and an oxygen–hydrogen single bond, making the keto tautomer less stable than the enol tautomer.
- Enol concentrations are low for most ketones (Figure 8.48).

Figure 8.48
Enols are rapidly converted back to the more stable keto tautomer.

- Resonance and hydrogen-bonding interactions can promote enol formation (Figure 8.49).

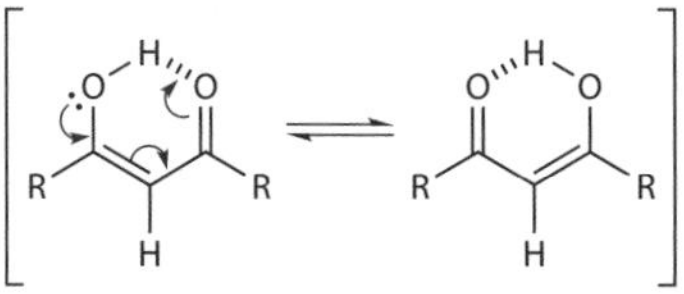

Figure 8.49
Enols from 1,3-dicarbonyl compounds are stable.

- Enols and enolates are nucleophilic and will react through carbon (Figure 8.50).

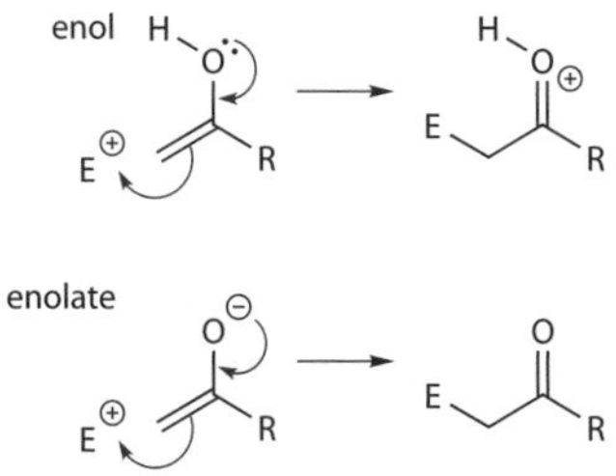

Figure 8.50
Generalised mechanism for the addition of electrophiles to enols and enolates.

- Overall the reaction substitutes one of the hydrogen atoms adjacent to the carbonyl with an electrophile.

Examples of α-hydrogen reactions

Alkylation reactions

- Although a wide range of acids and bases permit enol and enolate formation, to be synthetically useful the reaction conditions and reagents used must be carefully considered.
- Using aqueous sodium hydroxide will generate only a low concentration of enolate since under the reaction conditions the enolate can be converted back to the carbonyl. Sodium hydroxide is a reversible base.
    - When using a reversible base, the base and electrophile with which the enolate will react are added to the substrate at the same time.
    - Having a low concentration of enolate and high concentration of base means that, if additional α-hydrogen atoms are available in the product, these can also react after the first addition (Figure 8.51).

Tip

A reversible base can be used in reactions where a later step in the mechanism is irreversible; this channels the enolate towards product formation.

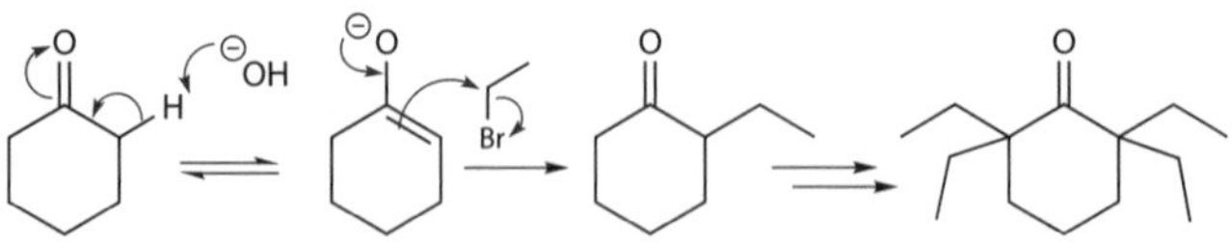

Figure 8.51
Enolate addition to an alkyl halide (reversible base).

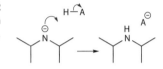

- The removal of subsequent α-hydrogens will generate more highly substituted and hence more stable enolates which are able to react with additional electrophiles. Stopping the reaction after one addition will be impossible.
- An irreversible base is a base which forms a conjugate acid which cannot be deprotonated under the reaction conditions.
- Irreversible bases include lithium diisopropylamide (LDA) and sodium hydride (Figure 8.52).

Figure 8.52
Lithium diisopropylamide is an irreversible base.

- Sodium hydride forms hydrogen gas which will be lost from the reaction mixture.
- An irreversible base can be used to convert all carbonyl molecules to the corresponding enolate at the same time; an electrophile can then be added to the reaction to deliver the addition product (Figure 8.53).
- Since the base and electrophile are being added sequentially only the monoaddition product is formed.

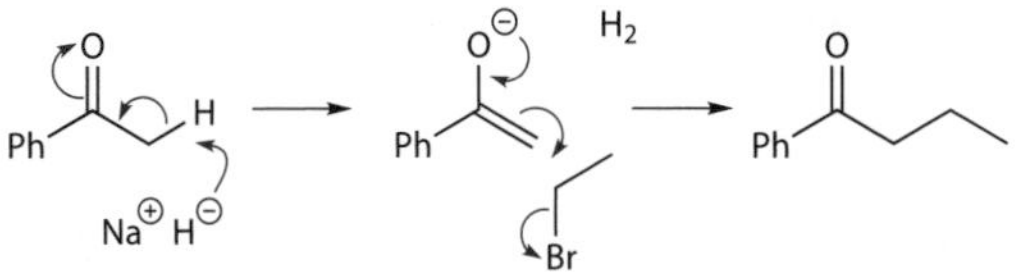

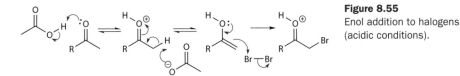

Figure 8.53
Enolate addition to an alkyl halide
(irreversible base).

Halogenation reactions (acidic conditions)

■ Under acidic reaction conditions carbonyl compounds can be
   halogenated at the α-position (Figure 8.54).

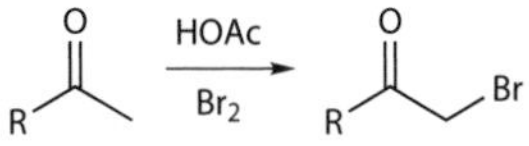

Figure 8.54
Halogenation under acidic conditions.

■ Enol formation is followed by nucleophilic attack on the
   bromine molecule.
■ The reaction is catalytic in acid as the acid is regenerated during
   the reaction (Figure 8.55).

Figure 8.55
Enol addition to halogens
(acidic conditions).

■ In principle the brominated product could react further since
   there are two additional enolisable α-hydrogen atoms available
   for substitution.
■ The presence of the electron-withdrawing bromine atom
   destabilises the positively charged transition state, making
   further deprotonation less favourable (Figure 8.56).

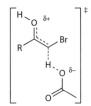

Figure 8.56
Transition state for enol formation.

■ Excess bromine and forcing reaction conditions are required for
   additional bromination at this carbon atom.

Halogenation reactions (basic conditions)

■ Under basic reaction conditions carbonyl compounds can also be
   halogenated at the α-position.
■ The product obtained is not the same as that formed under
   acidic conditions.

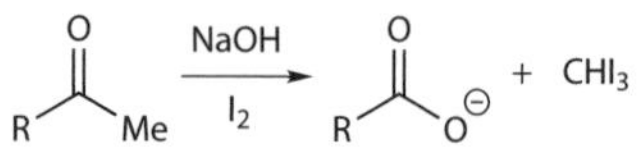

Figure 8.57
The iodoform reaction.

- The enolate formed under basic conditions is more reactive than the enol formed under acidic conditions and readily reacts with the iodine to form a mono-substituted carbonyl (Figure 8.57).
- The electron-withdrawing iodine increases the acidity of the two remaining α-hydrogen atoms through a weakening of the carbon–hydrogen σ bonds.
- The enolate formed from a subsequent deprotonation will experience a stabilising effect from the electron-withdrawing iodine, making this enolate more stable than the original enolate (Figure 8.58).
- Replacement of two hydrogen atoms by iodine will further increase the reactivity of the enolate and enable all three α-hydrogen atoms to be substituted.
- Having exchanged all three α-hydrogen atoms the molecule is now susceptible to nucleophilic attack by hydroxide, followed by elimination of CI_3^-, a good leaving group.
- Deprotonation of the carboxylic acid by CI_3^- generates iodoform and the corresponding carboxylate.

KeyPoints

Iodoform test
- Test for the presence of methyl ketones.
- Treating the ketone with iodine and aqueous sodium hydroxide results in the formation of a yellow precipitate of iodoform (CHI_3).

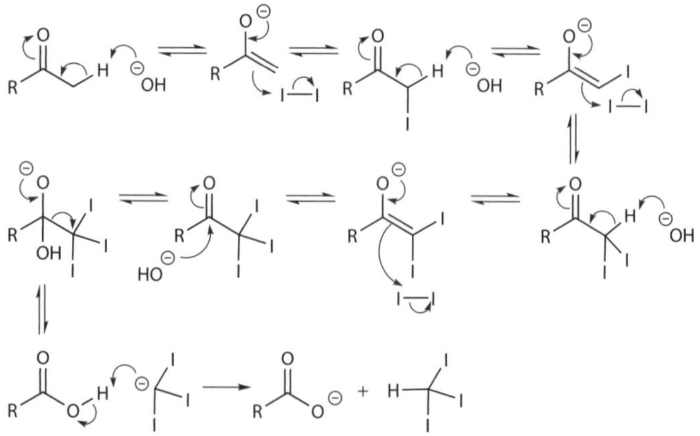

Figure 8.58
Enolate addition to halogens (basic conditions).

Aldol condensation (basic conditions)

- Since enolates are in equilibrium with the non-enolised carbonyl, in the absence of another electrophile, enolates will react with the non-enolised carbonyl in a nucleophilic addition reaction.
- The product contains both an aldehyde and an alcohol; hence the reaction is termed an aldol condensation (Figure 8.59).

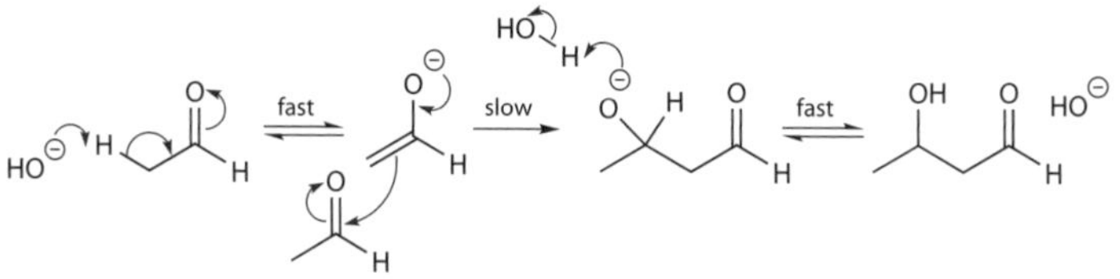

Figure 8.59
Enolate addition to an aldehyde (basic conditions).

- If the reaction is heated, the aldol product will undergo an E1cB elimination (elimination unimolecular conjugate base) to form an unsaturated aldehyde.
- In the E1cB reaction the leaving group is not lost from the starting substrate; it is lost from the conjugate base of the substrate (Figure 8.60).

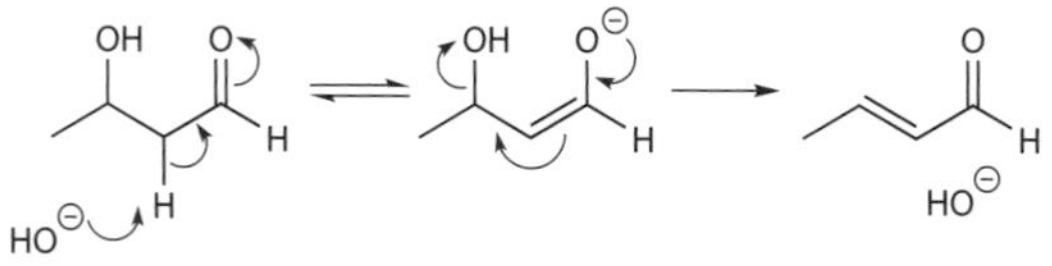

Figure 8.60
E1cB elimination (basic conditions).

Tips

- Enolate formation is rapid and reversible; the elimination of hydroxide is the rate-determining step and is irreversible.

- Hydroxide is a poor leaving group and will not undergo E2 reactions.
- Hydroxide will undergo E1cB elimination.

Aldol condensation (acidic conditions)

- Aldehydes will also undergo aldol condensations under acidic conditions if an alternative electrophile is not present (Figure 8.61).
- Initial protonation of the aldehyde generates the enol which reacts with a second molecule of protonated aldehyde to generate a new carbon–carbon bond in an irreversible addition.
- Under acidic conditions the aldol product is not stable and cannot be isolated.
- Protonation of the alcohol creates a good leaving group and an E1 reaction follows to generate an unsaturated aldehyde (Figure 8.62).

Tips

- Basic conditions – aldol product.
- Acidic conditions – elimination product (E1).
- Basic conditions with heat – elimination product E1cB).

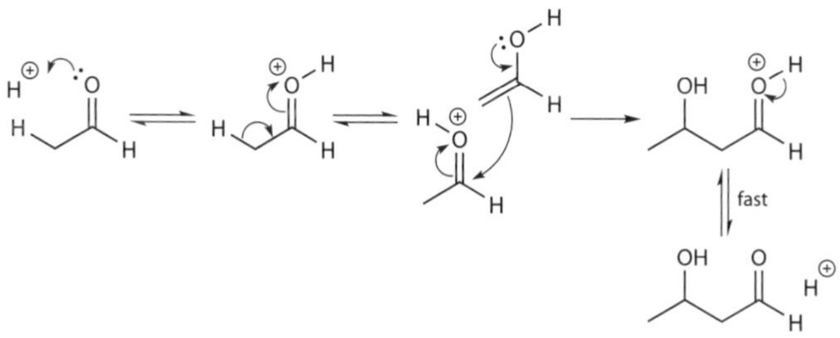

Figure 8.61
Enol addition to an aldehyde (acidic conditions).

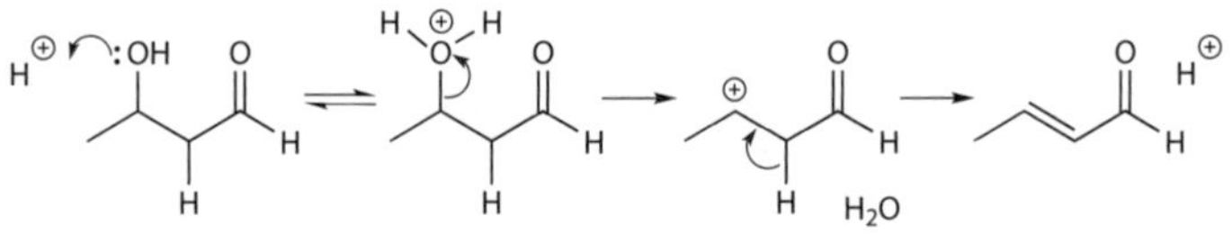

Figure 8.62
E1 elimination (acidic conditions).

Other aldol reactions

- Enols and enolates derived from ketones will also undergo aldol reaction with both aldehydes and other ketones.
- Reactions where the nucleophilic enol/enolate and electrophilic carbonyl are not derived from the same carbonyl parent are termed mixed aldol reactions (Figure 8.63).
  - Aldehydes with non-enolisable protons will always be the electrophilic partner in mixed aldol reactions (Figure 8.64).
  - Where there are two carbonyl groups in a molecule an intramolecular aldol condensation may occur if there is sufficient flexibility (Figure 8.65).
  - Enol/enolate formation at one carbonyl allows nucleophilic attack on the second electrophilic carbonyl and cyclises the molecule.

Tips

- The two protons a and b are not equivalent: deprotonation will generate two different enolates.
- Consider the two different aldol cyclisations individually.
- Six-membered rings will form in preference to four- or five-membered rings owing to reduced strain in the larger ring.

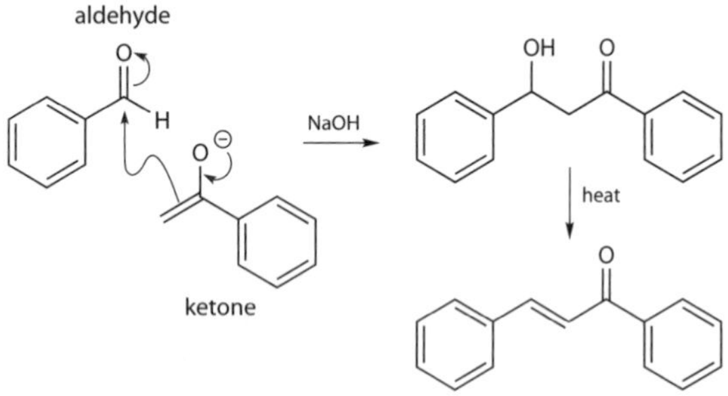

Figure 8.63
Mixed aldol reaction between a ketone enolate and an aldehyde.

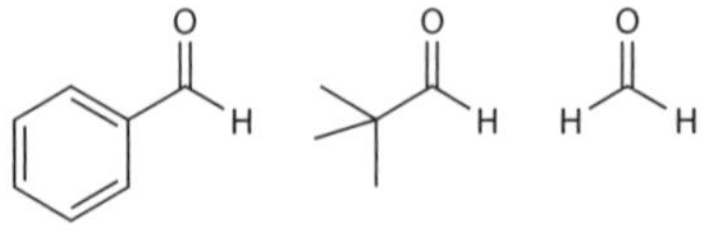

Figure 8.64
Non-enolisable aldehydes.

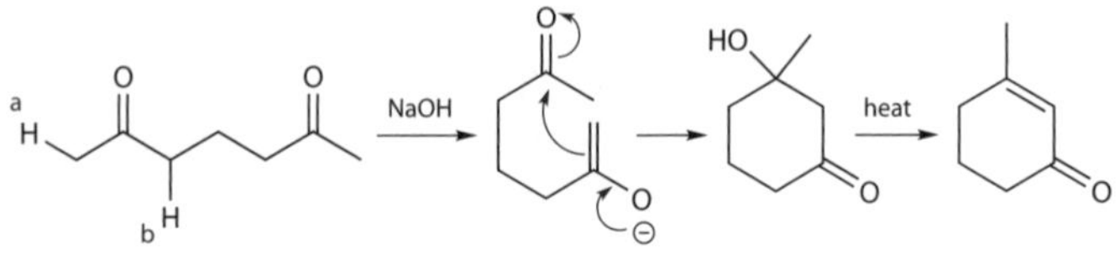

Figure 8.65
Intramolecular aldol condensation.

Claisen condensation

- Esters with acidic α-hydrogen atoms will form enolates under basic conditions.
- Ester enolates react in the same way as aldehyde- or ketone-derived enolates and will undergo self-condensation with non-enolised ester in the absence of an alternative electrophile; this is termed a Claisen condensation (Figure 8.66).
- Unlike the aldol condensation, the tetrahedral intermediate formed during a Claisen condensation has a leaving group attached to the carbon. A lone pair of electrons from the negatively charged oxygen push back to reform the π bond and eliminate an alkoxide anion.

KeyPoints

- It is important that the base, solvent and leaving group all 'match'. That is, if the ester is an ethyl ester, the base should be sodium ethoxide and the solvent ethanol.
- Mismatching would result in side product formation from addition–elimination reactions with the ester.
- By matching the groups, any addition–elimination reaction that occurs gives an identical product to the starting substrate.

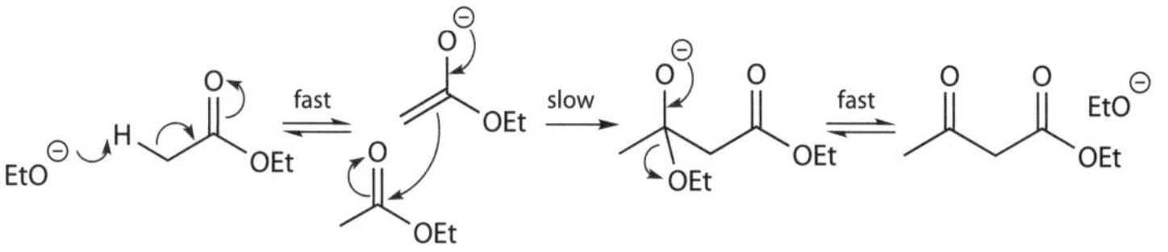

Figure 8.66 Ester enolate addition to an ester.

Dieckmann condensation

- When the two ester components of a Claisen condensation are in the same molecule, they react to form a cyclic keto-ester.
- An intramolecular Claisen condensation is called a Dieckmann condensation – a different name but exactly the same chemistry (Figure 8.67).

Tip

If you understand the chemistry of how molecules and reagents react, the name of a particular reaction or process is unimportant.

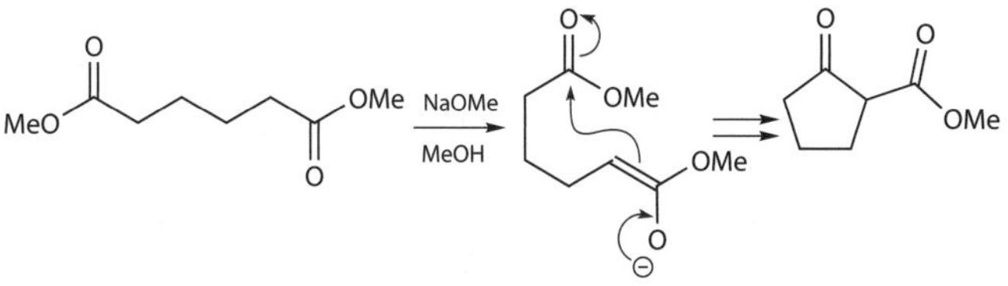

Figure 8.67 Intramolecular Claisen condensation (Dieckmann condensation).

Aldol and Claisen condensations in biology

- The aldol and Claisen condensations are two powerful methods to form carbon–carbon bonds; it should not be a surprise that biology uses an equivalent chemistry.

■ Nature makes use of a dimethylallyl pyrophosphate building block for the synthesis of a wide range of biologically important molecules ranging from vitamins to steroids (Figure 8.68).

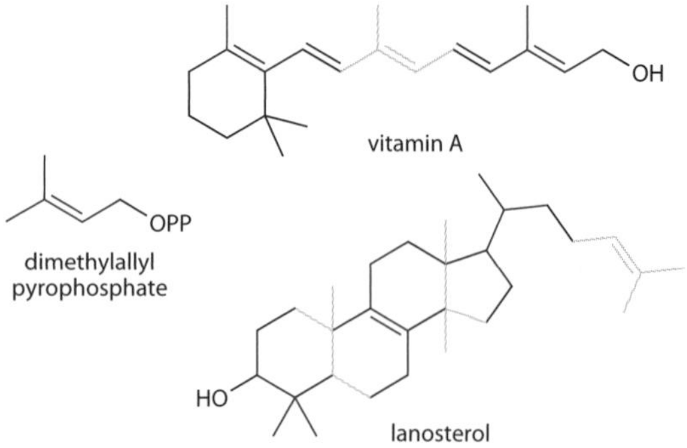

Figure 8.68
The dimethylallyl pyrophosphate building block.

vitamin A

dimethylallyl pyrophosphate

lanosterol

■ The dimethylallyl pyrophosphate building block is formed from the initial 'Claisen condensation' of acetyl coenzyme A to form acetoacetyl coenzyme A (Figure 8.69).

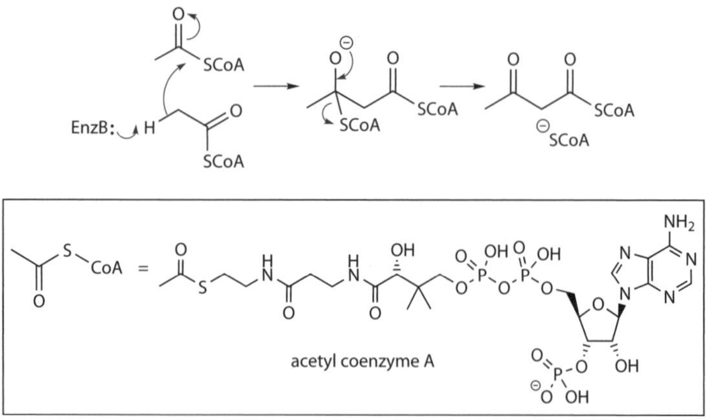

Figure 8.69
Simplified Claisen condensation of acetyl coenzyme A.

acetyl coenzyme A

■ The acetoacetyl coenzyme A then undergoes an aldol-like condensation with a third molecule of acetyl coenzyme A to assemble the carbon framework of the isoprene unit (Figure 8.70).

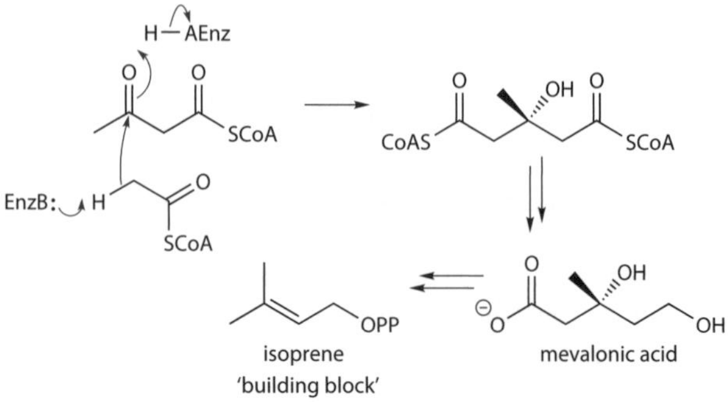

Figure 8.70
Simplified aldol condensation of acetoacetyl coenzyme A.

isoprene
'building block'

mevalonic acid

- A number of steps follow to modify individual carbon oxidation states but the isoprene core is already in place.

Conjugate addition (Michael addition)

- Nucleophiles attack the electrophilic carbonyl group at the electron-deficient carbon to undergo nucleophilic addition (Figure 8.71).
- If the carbonyl group is part of a conjugated system an alternative reaction pathway becomes available.
- The carbonyl group can now undergo resonance with the alkene, allowing electron density to be delocalised across the whole system (Figure 8.72).
- The delocalisation of electron density results in an alternating pattern of electron-rich and electron-deficient atoms.
- The terminal end of the conjugated system is electron-deficient and susceptible to attack by nucleophiles.

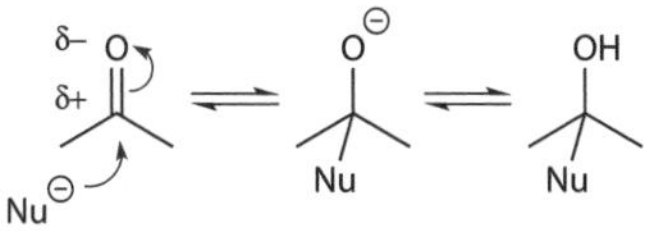

Figure 8.71
Generalised mechanism for nucleophilic addition.

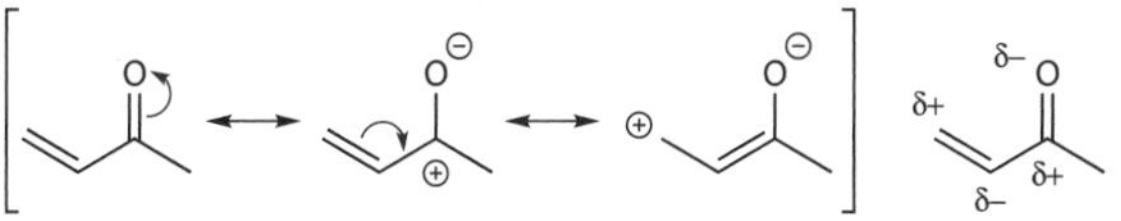

Figure 8.72
Resonance in conjugated carbonyl groups.

- When a nucleophile reacts with a conjugated carbonyl group the non-bonding electrons from the nucleophile attack the carbon–carbon π bond, forming a new σ bond between the nucleophile and the terminal carbon atom of the alkene (Figure 8.73).
- The pair of electrons from the alkene π bond form a new π bond to the carbonyl carbon and force the pair of electrons from the carbonyl π bond to retreat on to oxygen where they localise as a non-bonding pair.
- The intermediate enolate is then quenched through protonation during acidic workup to regenerate the carbonyl group.

Tip

Conjugate addition is sometimes referred to as 1,4-addition since the two ends of the reacting system are arranged in a 1,4-relationship.

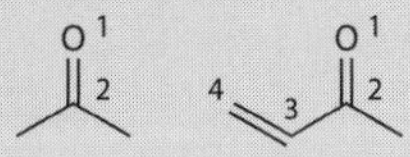

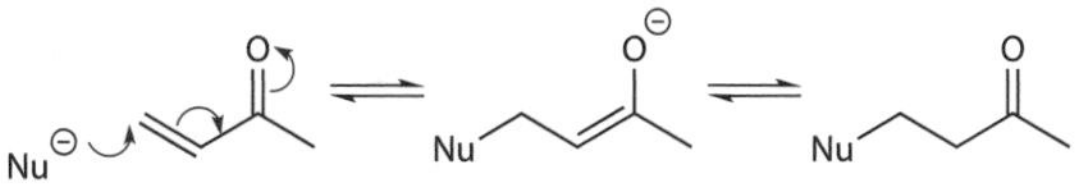

Figure 8.73
Generalised mechanism for conjugate addition.

Examples of conjugate addition

Etherification reactions

■ Good nucleophiles with a formal negative charge will participate in conjugate addition without the need for acid or base catalysis (Figure 8.74).

Figure 8.74
Conjugate addition of alkoxides.

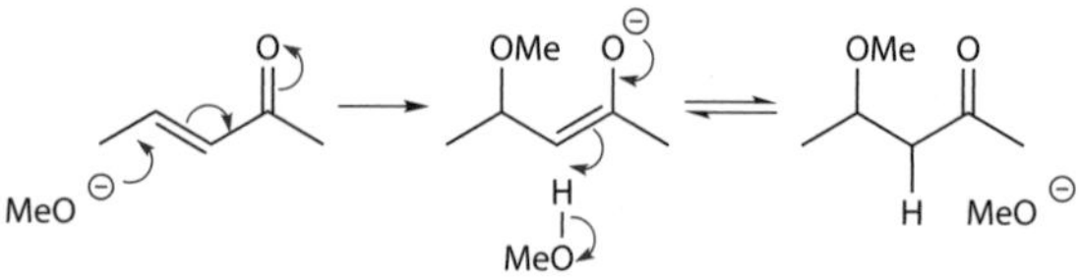

■ Conjugate addition can be thought of as the reverse of E1cB elimination (Figure 8.75).

Figure 8.75
E1cB elimination of methoxide.

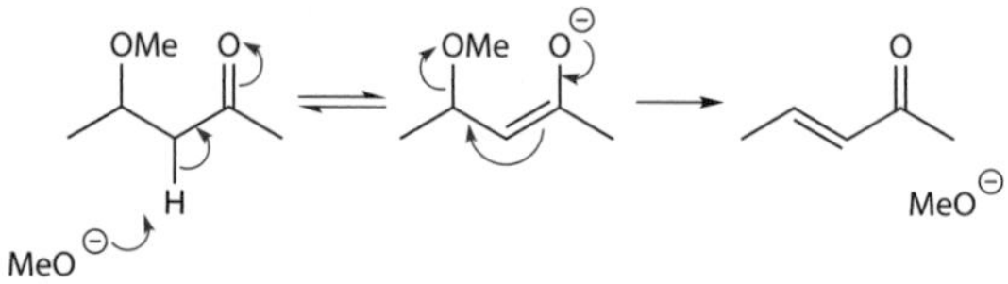

■ Under acidic reaction conditions, the electrophilicity of the conjugated system is increased, making reactions with less nucleophilic alcohols possible (Figure 8.76).

Figure 8.76
Conjugate addition of alcohols.

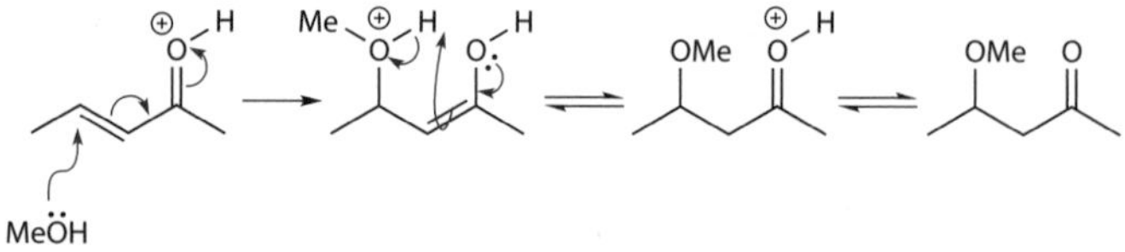

Reaction with amines and thiols

■ Amines and thiols are sufficiently nucleophilic to react with conjugated carbonyls without the need for catalysis (Figures 8.77 and 8.78).

Figure 8.77
Conjugate addition of amines.

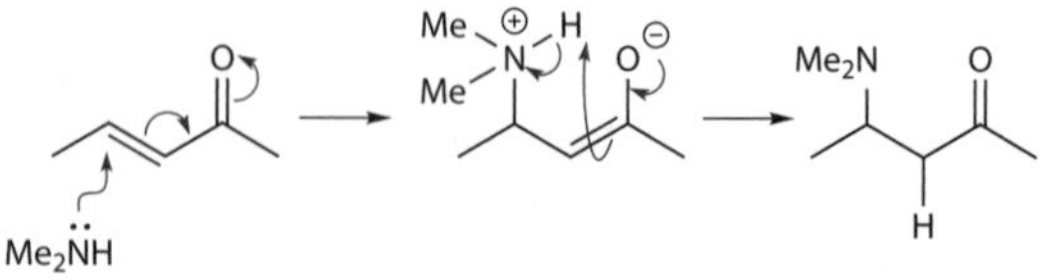

Figure 8.78
Conjugate addition of thiols.

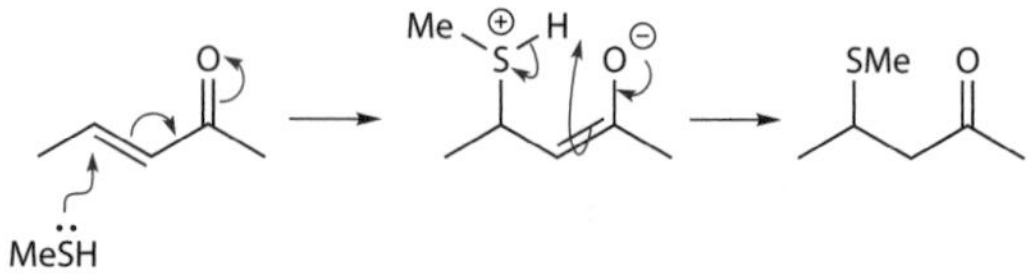

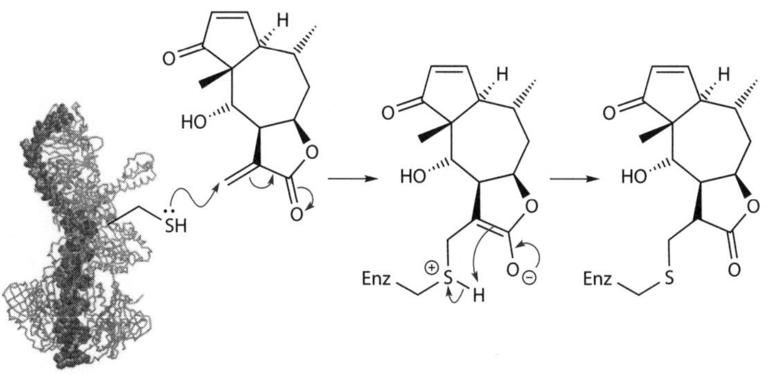

Figure 8.79
Conjugate addition
as a mechanism for
enzyme inhibition.

- The high nucleophilicity of thiols towards conjugated carbonyl groups has led to natural products, including helenalin, being investigated for the treatment of cancer.
- Helenalin is a powerful electrophile that undergoes conjugate addition with the thiol side chain of a cysteine residue within the active site of DNA polymerase.
- The helenalin molecule is irreversibly bound to the enzyme through the formation of a covalent bond, thereby inhibiting DNA polymerase and lowering nucleic acid synthesis (Figure 8.79).

Self-assessment

1. **Aldehydes readily undergo which of the following classes of reaction?**
a. electrophilic substitution
b. nucleophilic substitution
c. electrophilic addition
d. nucleophilic addition

2. **Identify one of the key intermediates in the following reaction:**

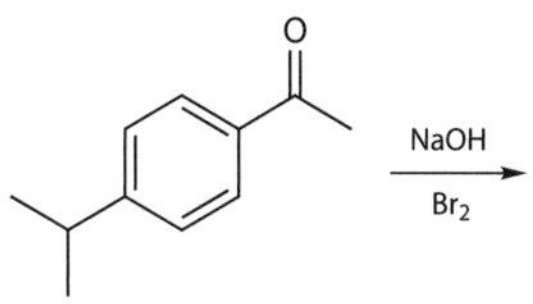

a.

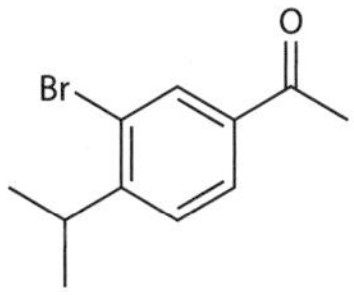

b.

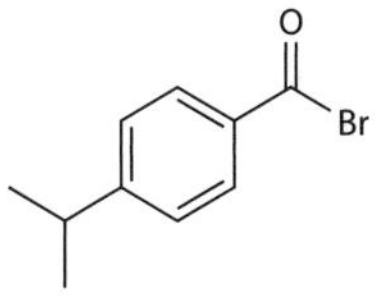

c.

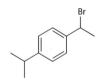

d.

Br

3. **Which of the following statements about the reactivity of esters is untrue?**
a. Nucleophiles always add reversibly to esters.
b. Esters are more reactive than amides.
c. Protonation of the carbonyl oxygen makes the ester more electrophilic.
d. Nucleophiles react with esters to give a tetrahedral intermediate.

4. **Which of the following amines cannot be used to form an imine?**
a.

b. NH₂

c. NH₂

d. NH₂

5. **Which of the following solvents would be unsuitable for a Grignard reaction?**
a. diethyl ether
b. dioxane
c. methanol
d. tetrahydrofuran

6. **Under acidic conditions two identical aldehydes undergo aldol condensation at room temperature to give what as the major product?**
a.

b.

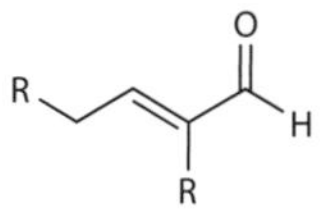

c.

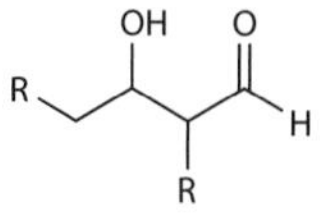

d.

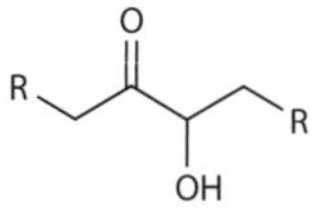

7. **Which of the following compounds is most reactive towards base-catalysed hydrolysis?**

a.

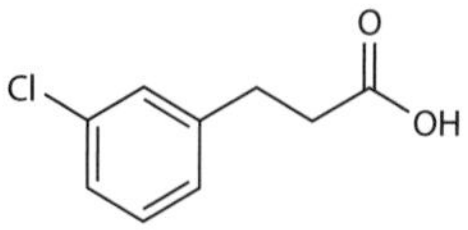

b.

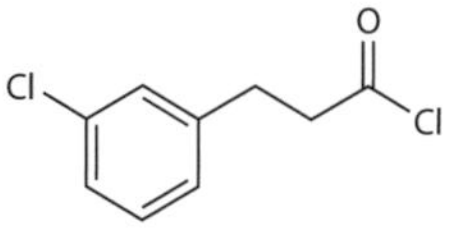

c.

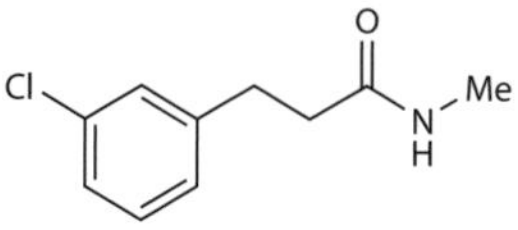

d.

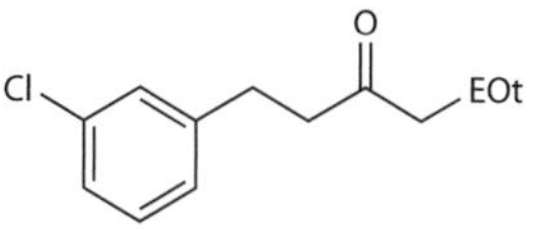

8. **Which of the following compounds will form the most stable hydrate?**

a.

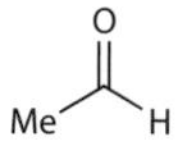

b.

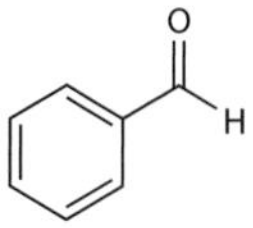

c.

d.

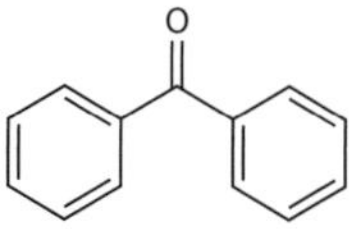

9. **Which of the following cannot exhibit keto–enol tautomerism?**

a.

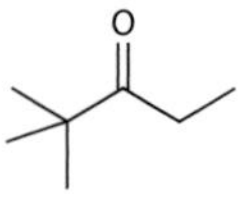

b.

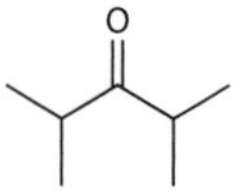

c.

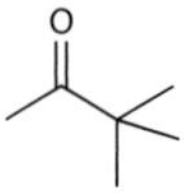

d.

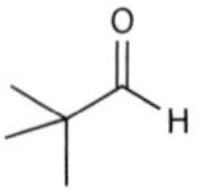

10. **Identify the major product of the following reaction:**

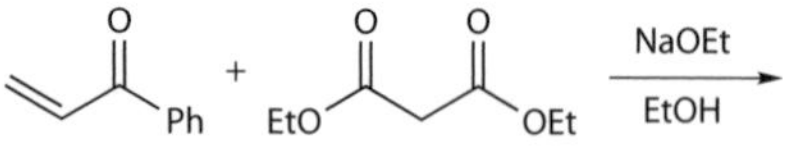

a.

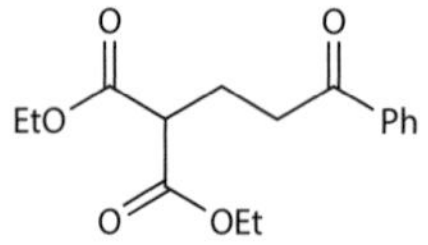

b.

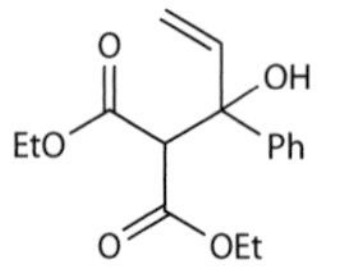

c.

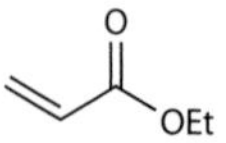

d.

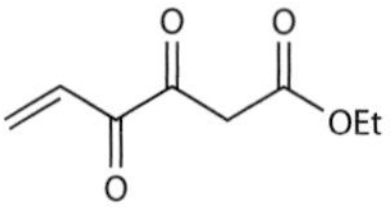

chapter 9

Chemistry of aromatic heterocyclic compounds

Overview

After learning the material presented in this chapter you should:
- be able to recognise aromaticity in molecules other than benzene
- appreciate how electron density is delocalised in heteroaromatic compounds and the implication of this for chemical reactivity
- be able to discuss the relationship between basicity and nitrogen hybridisation
- understand electrophilic substitution in heteroaromatic compounds
- understand nucleophilic substitution in pyridine
- appreciate the prevalence of heteroaromatic groups in pharmaceuticals.

Aromaticity in heterocyclic compounds

- Aromaticity is the stabilisation afforded to a molecule from the overlap of adjacent p orbitals in a cyclic, planar system which allows for the delocalisation of electron density across the whole molecule.
- Aromaticity will also occur for molecules containing atoms other than carbon.
- The requirement for 4n+2 electrons within the π system remains.
- Aromatic compounds that contain atoms other than carbon are termed aromatic heterocycles or heteroaromatic.
- Replacement of a C–H group in benzene with a nitrogen atom gives pyridine (Figure 9.1).
- To participate in resonance, the nitrogen atom on pyridine must be sp^2 hybridised (Figure 9.2).

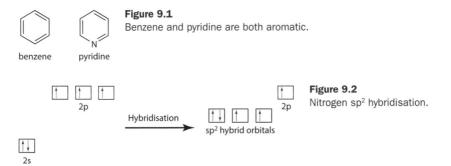

Figure 9.1
Benzene and pyridine are both aromatic.

benzene   pyridine

Figure 9.2
Nitrogen sp^2 hybridisation.

2p

Hybridisation

sp^2 hybrid orbitals

2p

2s

- Nitrogen has five valence electrons. Upon hybridisation there are two sp² hybrid orbitals containing a single electron; these form the σ bonds to the adjacent carbon atoms, making up the framework of the molecule.
- The pair of non-bonding electrons in the third sp² hybrid orbital is located in the plane of the ring.
- The single electron in the unhybridised p orbital is perpendicular to the plane of the ring and overlaps with the p orbitals on the adjacent carbon atoms (Figure 9.3).
- In five-membered heteroaromatic compounds the heteroatom will again be sp² hybridised to provide the necessary p orbital to allow electron delocalisation to take place (Figure 9.4).

Figure 9.3
Delocalisation of electron density in benzene and pyridine.

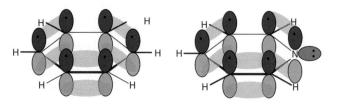

Figure 9.4
Delocalisation of electron density in pyrrole and furan.

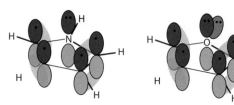

Naming of heteroaromatic compounds

- Heteroaromatic compounds are named after their parent structure (Figure 9.5).
- The positions on a pyridine ring relative to a functional group can be given as *ipso*, *ortho*, *meta* and *para* in the same way as benzene.
- The situation becomes more complicated when considering five-membered rings.

Figure 9.5
Common heteroaromatic parent compounds.

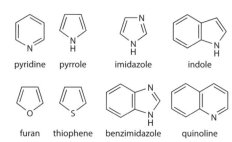

pyridine   pyrrole   imidazole   indole

furan   thiophene   benzimidazole   quinoline

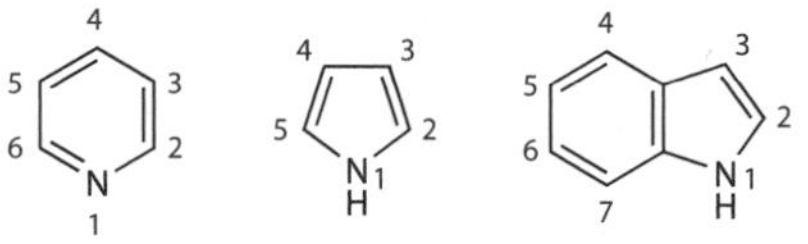

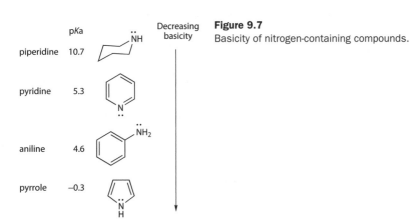

Figure 9.6 Numbering of heteroaromatic compounds.

- For heteroaromatic compounds it is more usual to consider the numerical positions on the ring (Figure 9.6).

The basicity of nitrogen heterocyclic compounds

- Not all nitrogen heteroaromatic compounds will be basic.
- When considering the basicity of amines it is necessary to take into account the hybridisation of the nitrogen atom and the location of the non-bonding pair of electrons within these hybrid orbitals (Figure 9.7).
- In piperidine the non-bonding pair of electrons is located in an sp³ hybridised orbital, whereas in pyridine the non-bonding pair of electrons is located in an sp² hybridised orbital.
- The greater s-character of an sp² hybridised orbital compared to an sp³ hybridised orbital means that the non-bonding pair of electrons on pyridine is more tightly held by the nitrogen nucleus.
- Pyridine is therefore less basic than piperidine, as the non-bonding electrons are less available for protonation.
- The nitrogen atom of aniline is sp² hybridised.
- Unlike pyridine, where the non-bonding pair of electrons is in the plane of the molecule, in aniline the non-bonding pair is located in a p orbital and is able to participate in resonance with the benzene ring (Figure 9.8).

Tips

- Substituents that are electron donating with benzene will be electron donating with heteroaromatic compounds.
- Substituents that are electron withdrawing with benzene will be electron donating with heteroaromatic compounds.
- The position(s) to which these groups direct may need to be re-evaluated in heteroaromatic compounds.

Figure 9.7
Basicity of nitrogen-containing compounds.

	pKa	
piperidine	10.7	
pyridine	5.3	
aniline	4.6	
pyrrole	−0.3	

Decreasing basicity

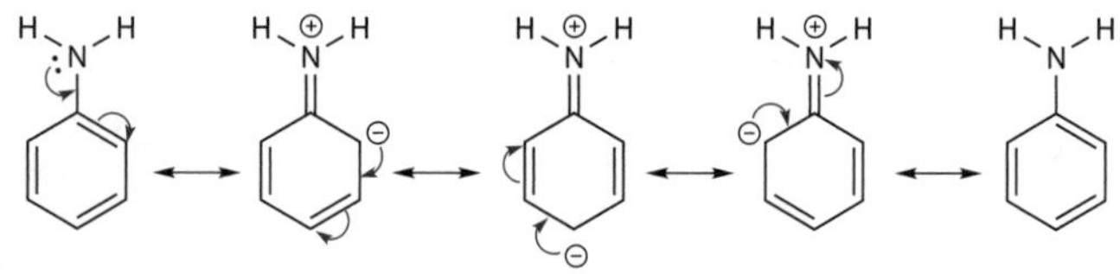

Figure 9.8
Delocalisation of electron
density in aniline.

The delocalisation of the non-bonding electrons in aniline makes the electrons less available for protonation. As a result aniline will be less basic than pyridine, despite the nitrogen atoms in both molecules being sp² hybridised.

The nitrogen atom of pyrrole is sp² hybridised.

In the case of pyrrole the non-bonding pair of electrons is located in a p orbital which is contributing to the aromaticity of the molecule.

Protonation of the pyrrole nitrogen would break the aromatic stabilisation of the molecule as there would only be four electrons contributing to the π system.

KeyPoints

- Although a molecule may have a 'non-bonding' pair of electrons, it is important to consider whether the electrons are available to participate in chemical reactions.

- Pyrrole is non-basic; the lone pair of electrons on the nitrogen is not available for protonation (Figure 9.9).

Figure 9.9
Comparison of pyridine and pyrrole electronic structure.

Pyridine – Basic

Pyrrole – Non-basic

Electrons in sp² hybrid orbital not involved in resonance.

Electrons in a 2p orbital are involved in resonance.

Reactivity of heteroaromatic compounds

Pyridine

- In pyridine the non-bonding pair of electrons on nitrogen is located in an sp² hybrid orbital and cannot participate in resonance (Figure 9.10). If curly arrows are drawn, the resulting resonance structure is clearly incorrect.
- The presence of the electronegative nitrogen atom will inductively withdraw electron density through the σ-bond framework of pyridine.

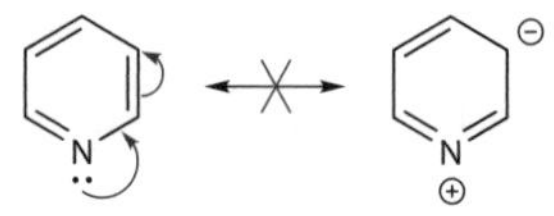

Figure 9.10
Delocalisation of the sp² hybridised electron pair in pyridine.

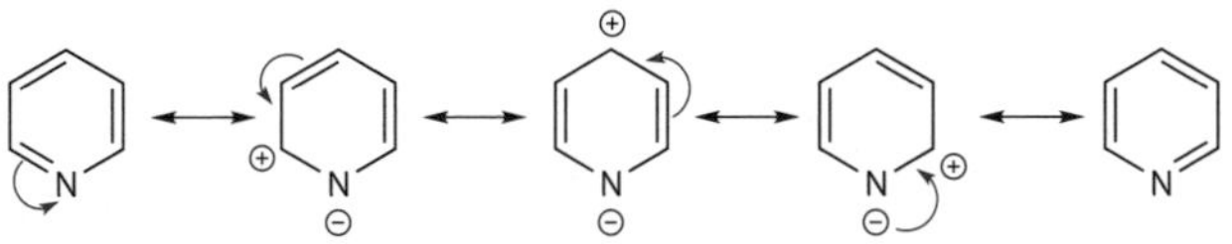

Figure 9.11
Delocalisation of electron density in pyridine.

- Pyridine will undergo resonance through the delocalisation of the electron in the p orbital (Figure 9.11).
- The higher electronegativity of nitrogen relative to carbon will stabilise those resonance structures where there is a negative charge located on the nitrogen atom.
- The pyridine ring is electron-deficient.

Pyrrole and furan

- In pyrrole there is an inductive withdrawal of electron density towards the electronegative nitrogen atom.
- The pair of non-bonding electrons in the p orbital on nitrogen is available for resonance and can be delocalised into the ring system.
- Electron delocalisation is not uniform across the entire ring: electron density is highest at the 2- and 3-positions.
- Pyrrole is nucleophilic at the 2- and 3-positions.
- The donation of electron density into the ring system makes pyrrole electron-rich (Figure 9.12).
- Furan is able to donate electron density into the aromatic system in the same way as pyrrole (Figure 9.13).
- Due to the higher electronegativity of oxygen compared with nitrogen there will be a reduced contribution of those resonance hybrids bearing a positive charge on the oxygen atom.

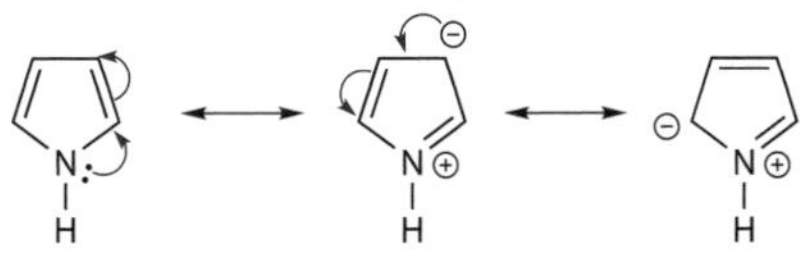

Figure 9.12
Delocalisation of electron density in pyrrole.

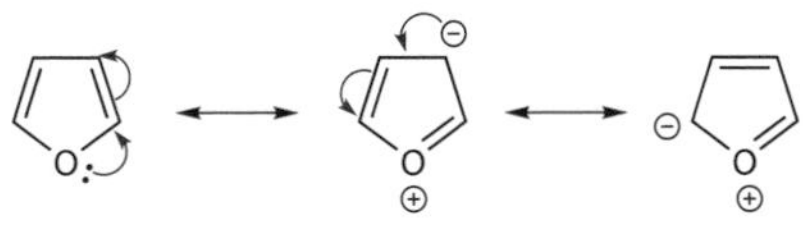

Figure 9.13
Delocalisation of electron density in furan.

Indole

- In indole the non-bonding pair of electrons in the p orbital on nitrogen is available for resonance and can be delocalised into the ring system.

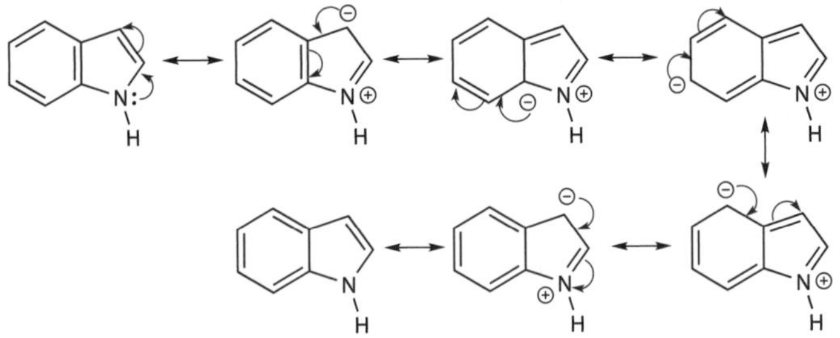

Figure 9.14
Delocalisation of electron density in indole.

- Electron delocalisation is not uniform across the entire ring system; the individual resonance hybrids do not contribute equally.
- Those resonance hybrids where the aromaticity in both rings is broken contribute to a lesser extent.
- Indole will be nucleophilic at the 3-position.
- The donation of electron density into the ring system makes indole electron-rich (Figure 9.14).

Imidazole

- In imidazole the non-bonding pair of electrons in the p orbital of one sp² hybridised nitrogen is available for resonance and can be delocalised into the ring system. The non-bonding electrons on the other nitrogen are in an sp² hybrid orbital and are unavailable.
- Electron delocalisation is not uniform across the entire ring system: the individual resonance hybrids do not contribute equally (Figure 9.15).

Figure 9.15
Delocalisation of electron density in imidazole.

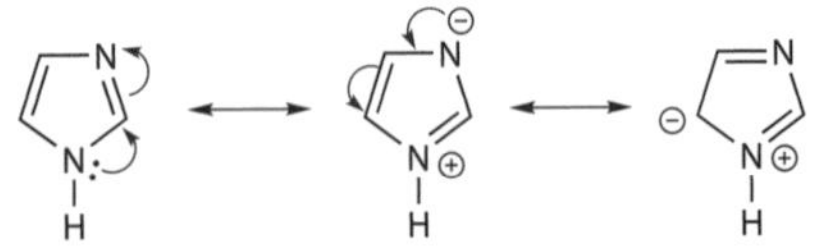

Electrophilic substitution in pyridine

- Unlike benzene, pyridine will not undergo electrophilic substitution reactions due to the electron-deficient nature of the ring system which greatly reduces the nucleophilicity of pyridine.
- Addition of an electrophile at the 2- or 4-position of pyridine would also result in a destabilised carbocation intermediate (Figure 9.16).
- The non-bonding pair of electrons on nitrogen is available to react as a nucleophile.
- Under the reaction conditions for electrophilic substitution there would therefore be competition between the ring and the nitrogen atom for the electrophile.

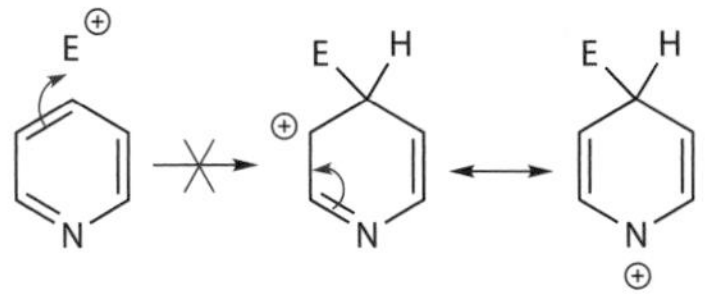

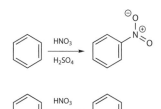

Figure 9.16
Electrophilic substitution in pyridine.

■ If subjected to the standard nitration conditions, pyridine would not undergo electrophilic substitution. Instead, under the highly acidic reaction conditions, the nitrogen atom would be protonated (Figure 9.17).

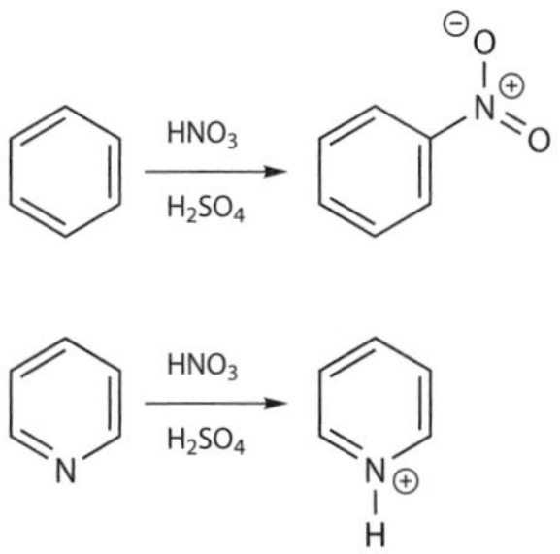

Figure 9.17
Nitration of benzene versus nitration of pyridine.

Electrophilic substitution with an activating group

■ Activated pyridines will undergo electrophilic substitution.
■ Amino and methoxy groups are able to donate electron density to the pyridine ring to increase the nucleophilicity of pyridine (Figure 9.18).
■ The activating group will direct electrophilic substitution as it would in benzene. That is, *ortho*- and *para*-addition products will dominate (Figure 9.19).

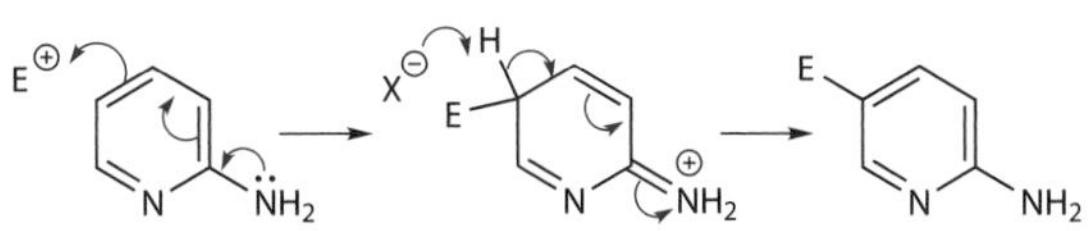

Figure 9.18
Electrophilic substitution in activated pyridines.

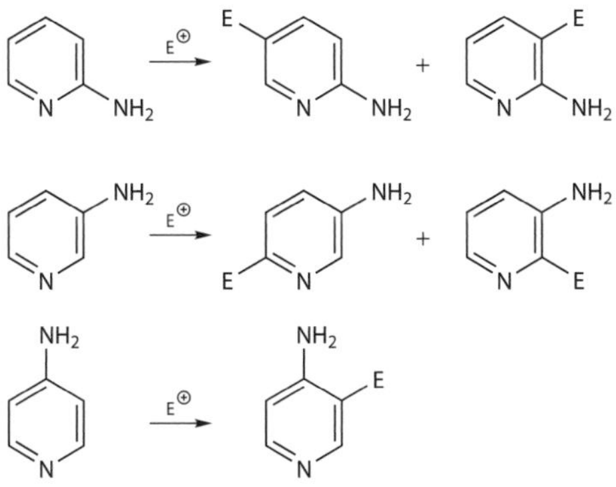

Figure 9.19
Substitution patterns for electrophilic substitution in pyridine.

Electrophilic substitution in pyrrole

- Both pyrrole and furan undergo electrophilic substitution.
- The electron-rich nature of these molecules will make them more powerful nucleophiles than unactivated benzene.
- Electrophilic substitution can occur at either the 2- or 3-position (Figure 9.20).
- The major substitution product corresponds to addition of the electrophile at the 2-position.
- The cationic intermediate formed from addition at the 2-position is more stable than that formed from addition at the 3-position due to the positively charged nitrogen being conjugated with both double bonds at the same time.

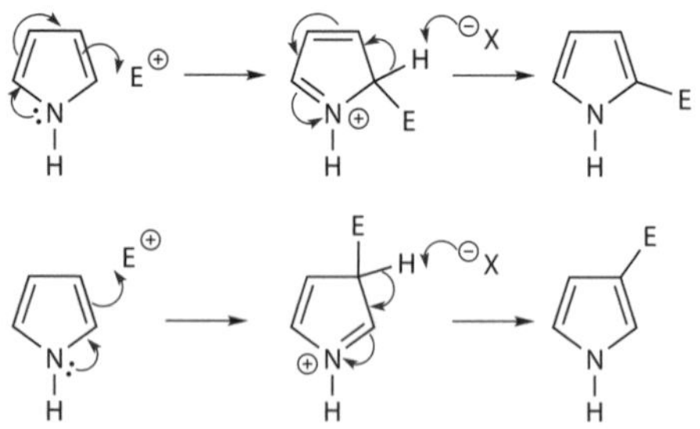

Figure 9.20
Electrophilic substitution in pyrrole.

KeyPoint

- Electrophilic substitution at the 2-position of indole would require the aromaticity of both rings to be broken, which would be highly destabilising.

Electrophilic substitution in indole

- Indole will undergo electrophilic substitution as it is electron-rich and therefore a good nucleophile.
- Electrophilic substitution will occur exclusively at the 3-position (Figure 9.21).

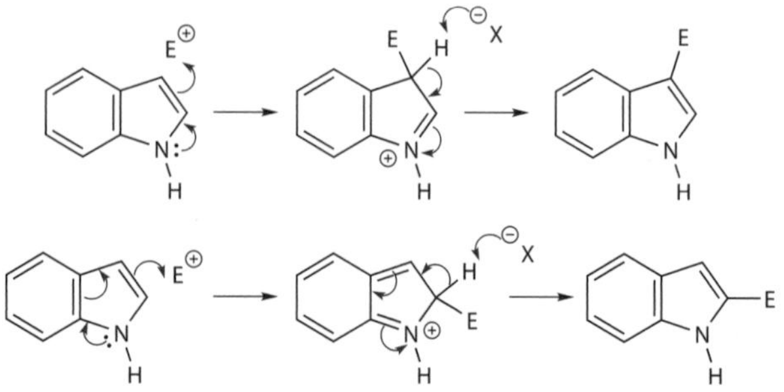

Figure 9.21
Electrophilic substitution in indole.

Substituted heteroaromatic derivatives

■ If a substituent is present on the heteroaromatic compound it will affect both the rate of addition of an electrophile and where the electrophile adds.

■ It is necessary to consider each intermediate individually and evaluate the stability of each by comparing resonance stabilisation.

■ Electron-donating groups will activate the heteroaromatic compound towards electrophilic substitution.

■ Electron-withdrawing groups will deactivate the heteroaromatic compound towards electrophilic substitution.

Nucleophilic substitution

■ Although pyridine will not undergo electrophilic substitution, the electron-deficient nature of the molecule will make it more reactive towards nucleophiles.

■ Pyridine will react with nucleophiles in nucleophilic substitution reactions.

■ For nucleophilic substitution to take place in pyridine there must be a viable leaving group present at either the 2- or 4-position (Figure 9.22).

■ When a nucleophile attacks pyridine the non-bonding electrons from the nucleophile attack one of the π bonds, forming a new σ bond between the nucleophile and the carbon atom.

■ Nucleophilic attack can occur through either '1,2-addition' or '1,4-addition'.

■ The process of breaking the π bond results in the movement of a pair of electrons on to the nitrogen atom.

■ The intermediate anion is stabilised both by the electronegative nitrogen atom and through delocalisation around the pyridine ring.

■ Since there is a leaving group on the carbon to which the nucleophile is attached, the tetrahedral intermediate will collapse. A non-bonding pair of electrons from the nitrogen will reform the π system and displace the electrons from the σ bond on to the leaving group.

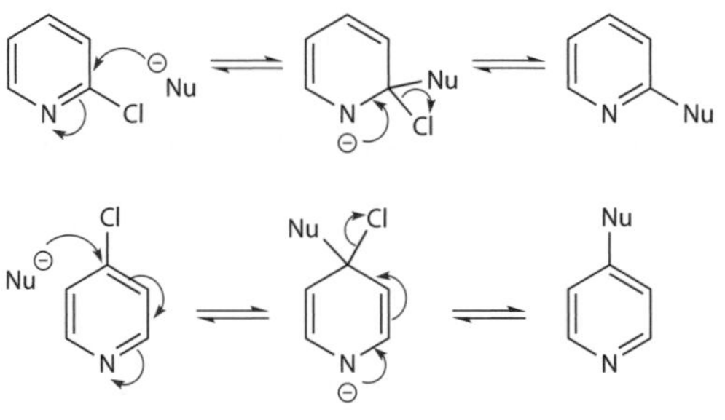

Figure 9.22
Generalised mechanism for nucleophilic aromatic substitution.

Figure 9.23
Reaction profile for
nucleophilic aromatic
substitution.

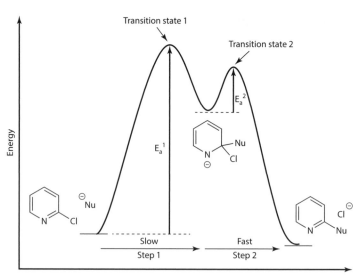

KeyPoints

- Activated pyridines undergo electrophilic substitution.
- Unactivated pyridines undergo nucleophilic substitution.

■ Benzene substituted with one or more strongly electron-withdrawing groups can be forced to undergo nucleophilic substitution instead of the more usual electrophilic substitution (Figure 9.24).

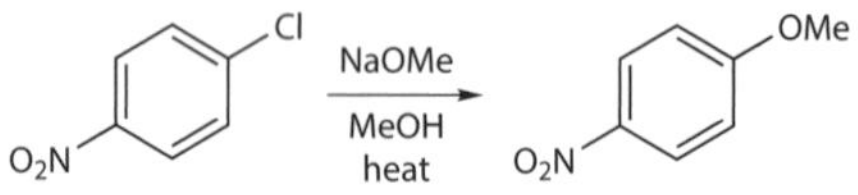

Figure 9.24
Nucleophilic substitution occurs in electron-deficient benzene derivatives.

Examples of heteroaromatic drugs

■ Heteroaromatic groups are found in a diverse range of pharmacologically active molecules, both those that are naturally occurring and those that have been developed by the pharmaceutical industry.

■ Many of the biggest-selling drugs currently on the market contain at least one heteroaromatic group (Figure 9.25).

Reasons for the prevalence of heteroaromatic groups in pharmaceuticals

■ The presence of the heteroatom(s) allows the heteroaromatic group to participate in hydrogen-bonding interactions with the target protein, increasing the affinity of the molecule towards the protein when compared to the all-carbon analogue.

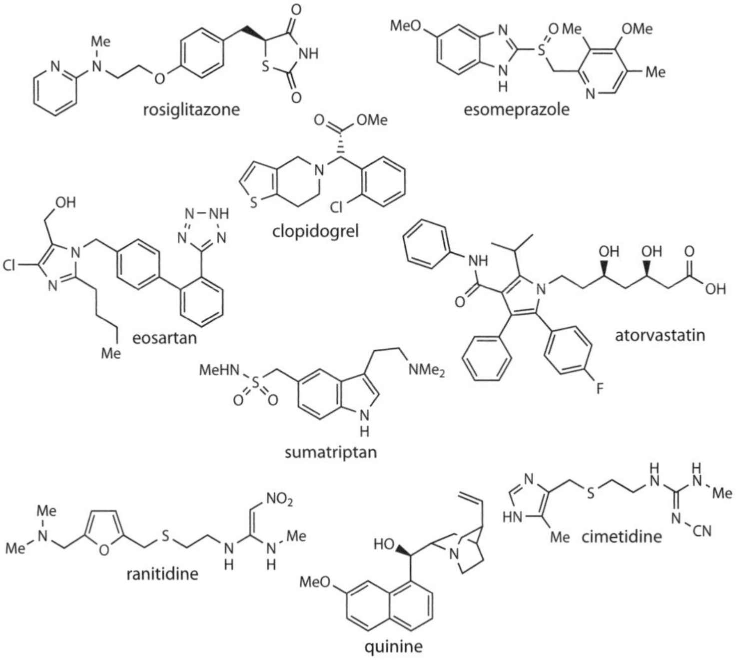

Figure 9.25 Drugs containing heteroaromatic groups.

- The synthesis of heteroaromatic groups is straightforward and well documented, with many of the necessary building blocks being commercially available.
- It is possible to modify and fine-tune the properties of a molecule either by changing the substituents on a heteroaromatic group or by changing to an alternate heteroaromatic group.
- Heteroaromatic groups demonstrate good *in vivo* stability.
- 'Scaffold hopping' from one heteroaromatic group to another can increase the novelty of a molecule; there is high commercial value in being able to do this as it allows for patent protection of new discoveries.

Self-assessment

1. **Which of the following substituted pyridines will be the most basic?**

a.

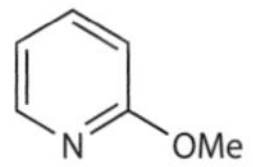

b.

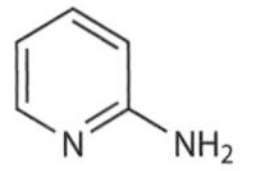

c.

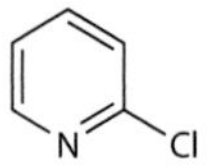

d.

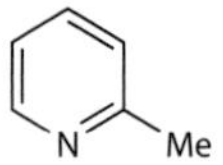

2. Identify the major product of the following reaction:

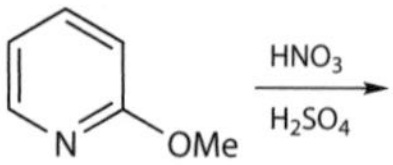

a.

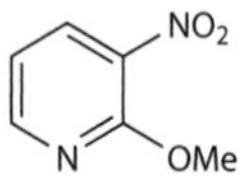

b.

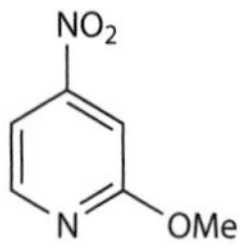

c.

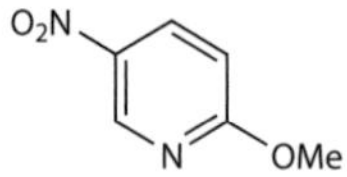

d.

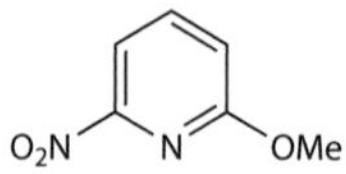

3. Which of the following heteroaromatic compounds will be the most basic?

a.

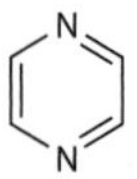

b.

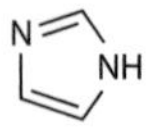

c.

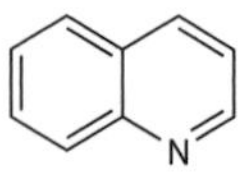

d.

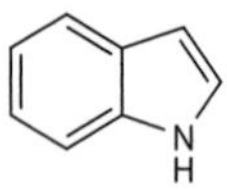

4. Which of the following is not a valid resonance hybrid for indole?

a.

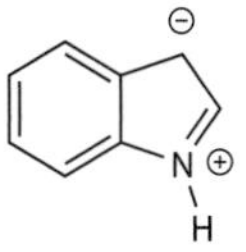

b.

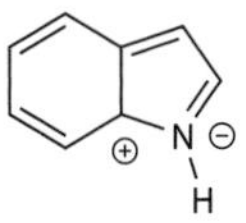

c.

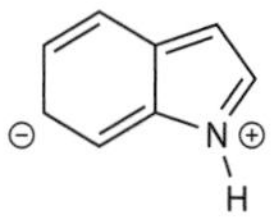

d.

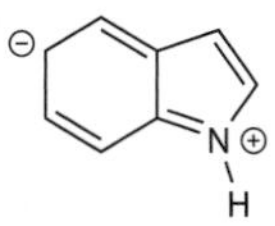

5. Which of the following statements concerning the reactivity of benzene and
    pyrrole towards electrophiles is correct?
a. Pyrrole is unreactive towards electrophiles.
b. Pyrrole is less reactive towards electrophiles than benzene.
c. Pyrrole is more reactive towards electrophiles than benzene.
d. Pyrrole and benzene have comparable reactivities towards electrophiles.

6. Identify the major product of the following reaction:

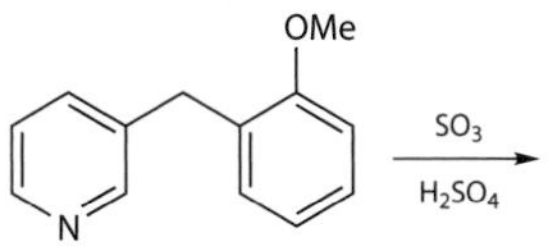

a.

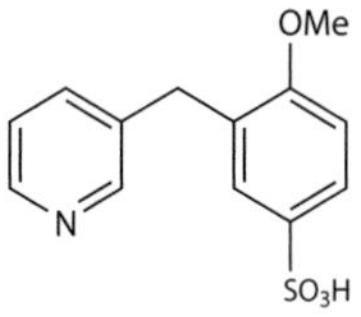

b.

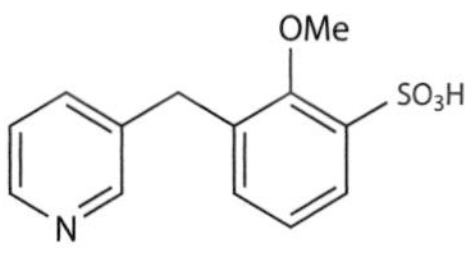

c.

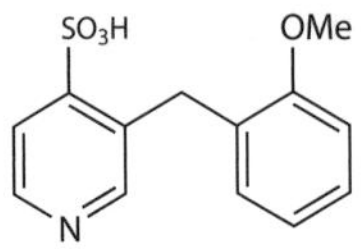

d.

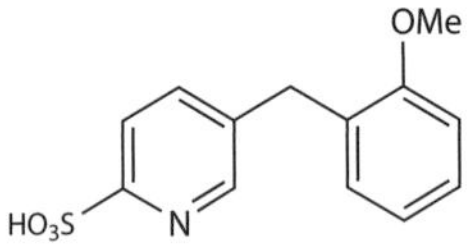

7. **The following is an example of what class of reaction?**

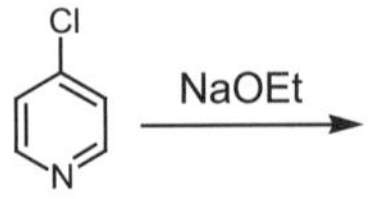

a. electrophilic addition
b. electrophilic substitution
c. nucleophilic addition
d. nucleophilic substitution

8. **Which of the following statements correctly describes the distribution of non-bonding electrons on the oxygen atom of furan?**
a. a pair of electrons in an sp^2 hybrid orbital and a pair of electrons in a p orbital
b. two pairs of electrons in sp^3 hybrid orbitals
c. a pair of electrons in an sp^3 hybrid orbital and a pair of electrons in a p orbital
d. two pairs of electrons in perpendicular p orbitals

9. **Identify the major product of the following reaction:**

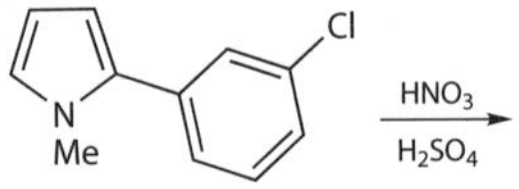

a.

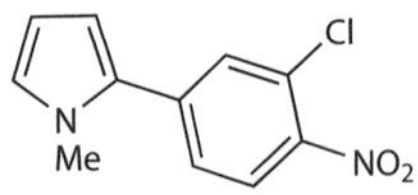

b.

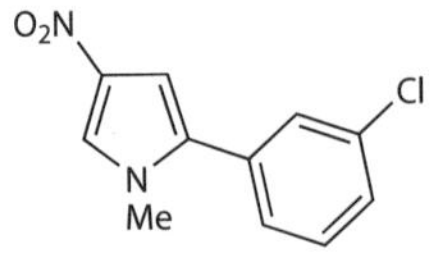

c.

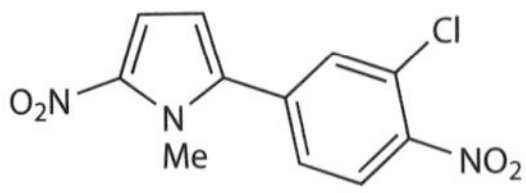

d.

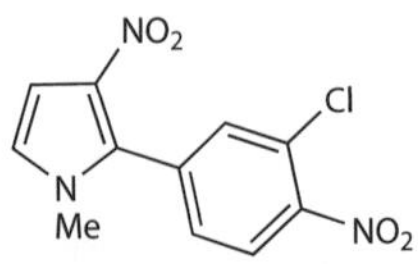

10. **Identify the major product of the following reaction:**

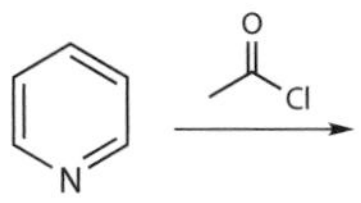

a.

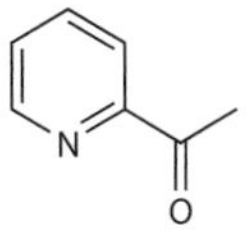

b.

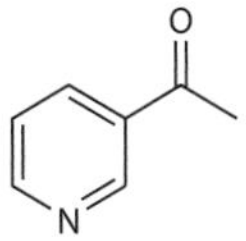

c.

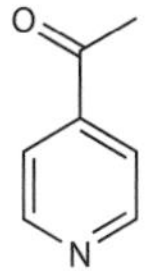

d.

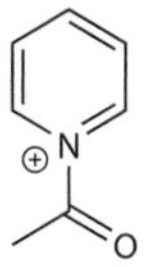

chapter 10

Amino acids, peptides and proteins

Overview

After learning the material presented in this chapter you should:

- recognise and be able to name and draw the 20 different amino acids found in proteins
- be able to identify the component amino acids in a peptide given its peptide sequence/primary structure
- know the different types of protein secondary structures: α-helices, β-pleated sheets and reverse turns
- understand how proteins fold and the forces that contribute to their stability
- understand why some proteins exhibit a quaternary structure
- understand the differences in structure between globular and fibrous proteins, and between soluble proteins and membrane proteins
- understand how environmental stresses lead to protein denaturation
- be familiar with the analytical methods employed (a) to test for amino acids and (b) to assay for protein.

Amino acids

- All proteins and peptides found in humans are linear (unbranched) polymers of L-α-amino acids. α-amino acids have the general structural formula shown in Figure 10.1.
- In this formula R is referred to as the amino acid *side chain*, and is a substituent group that varies from one amino acid to another. The prefix α indicates that both the amino and carboxyl groups are attached to the same (alpha) carbon atom.
- There are 20 different protein amino acids, which for convenience are subdivided according to the relative polarity of their side chains (Table 10.1).
- The apolar amino acids include those with aliphatic side chains, like alanine and valine, and those with aromatic side chains, like phenylalanine and tyrosine.
- The uncharged polar amino acids, like serine and asparagine, have side chains that bear non-ionisable groups that are able to hydrogen bond to water.
- The charged polar amino acids, like lysine and glutamic acid, have side chains that bear ionisable groups – either cationic (lysine) or anionic (glutamic acid).

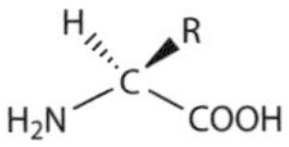

Figure 10.1
Structure of α-amino acids.

Tip

Amino acids can be detected/ visualised by reacting with *ninhydrin*. A deep blue or purple colour results when the amino acids react with the ninhydrin. This can be useful in the development of amino acid spots separated in paper or thin-layer chromatography, and it is also the means by which forensic scientists reveal latent fingerprints – the ninhydrin here reacts with amino acids found in sweat deposited from the unique pattern of ridges on the fingertips.

Tip

D-amino acids are sometimes used in making peptide drugs – with the aim of increasing their half-life in the body – slowing down their degradation by the body's proteolytic enzymes (which have evolved to deal specifically with peptides and proteins built from L-amino acids).

- Proline is unique among the protein amino acids in that it is actually an α-*imino acid*.
- The amino acid cysteine has a side chain that bears a thiol group and this can be oxidised in reaction with another cysteine side chain to form the amino acid cystine, with the two amino acid side chains covalently linked through a *disulfide bond* (sometimes also referred to as a disulfide link(age)).
- All human proteins and peptides are based on L-α-amino acids; D-amino acids are sometimes encountered in the peptides and proteins produced in organisms like bacteria and fungi (Figure 10.2).
- For convenience in summarising the covalent structures of peptides and proteins, the details of their constituent amino acids are often presented using amino acid three-letter or one-letter codes (Table 10.1).

Peptides

- Polymerisation of amino acids to form peptides and proteins occurs through a condensation reaction between the α-amino group of one amino acid and the α-carboxyl group of another. The amino acids thus become bonded together through the formation of secondary amide bonds, more commonly referred to as *peptide bonds* (Figure 10.3).

Figure 10.2
Configuration of L- and D-amino acids.

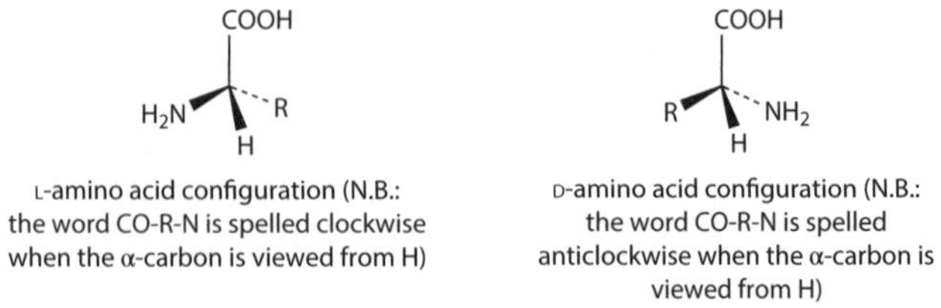

L-amino acid configuration (N.B.: the word CO-R-N is spelled clockwise when the α-carbon is viewed from H)

D-amino acid configuration (N.B.: the word CO-R-N is spelled anticlockwise when the α-carbon is viewed from H)

Figure 10.3
Peptide bonds.

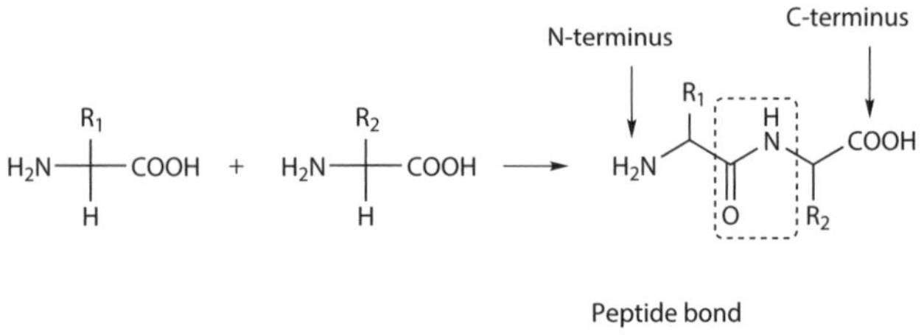

Peptide bond

Table 10.1 Protein amino acid side chains and three- and one-letter codes

Amino acid	Side chain (R group)	3- (and 1-) letter codes	Amino acid	Side chain (R group)	3- (and 1-) letter codes
Alanine	$-CH_3$	Ala (A)	Lysine	$-(CH_2)_4NH_2$	Lys (K)
Arginine	$-(CH_2)_3NH-C(=NH)NH_2$	Arg (R)	Methionine	$-CH_2CH_2SCH_3$	Met (M)
Tyrosine	$-CH_2-C_6H_4-OH$	Tyr (Y)	Phenylalanine	$-CH_2-C_6H_5$	Phe (F)
Asparagine	$-CH_2CONH_2$	Asn (N)	Proline	(ring structure, HN—CH—COOH)	Pro (P)
Cysteine	$-CH_2SH$	Cys (C)	Serine	$-CH_2OH$	Ser (S)
Glutamic acid	$-CH_2CH_2COOH$	Glu (E)	Threonine	$-CH(CH_3)OH$	Thr (T)
Histidine	$-CH_2$ (imidazole ring)	His (H)	Tryptophan	$-H_2C$ (indole ring)	Trp (W)
Glycine	$-H$	Gly (G)	Glutamine	$-CH_2CH_2CONH_2$	Gln (Q)
Aspartic acid	$-CH_2COOH$	Asp (D)	Leucine	$-CH_2CH(CH_3)_2$	Leu (L)
Isoleucine	$-CH(CH_3)CH_2CH_3$	Ile (I)	Valine	$-CH(CH_3)_2$	Val (V)

Figure 10.4

Peptide bonds. These bonds show significant double bond character, existing as a hybrid of the charged and uncharged resonance forms (upper panel). There is thus restricted rotation about the N–C bond, and *cis* and *trans* forms exist (lower panel).

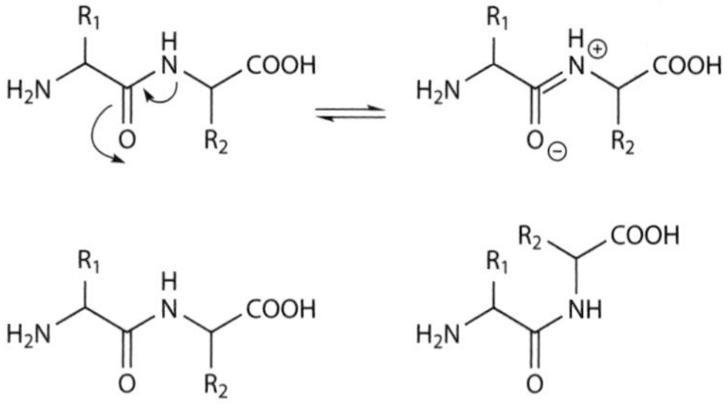

trans-peptide bond   *cis*-peptide bond

Tip

Aspartame, the α-carboxy methyl ester of the dipeptide Asp-Phe (shown in Figure 10.5), is a pharmaceutical excipient commonly used as an artificial sweetener. Enzymatic hydrolysis of this compound to its component amino acids in the small intestine may present a problem for patients suffering from phenylketonuria, a rare inherited condition in which the patient is unable to metabolise phenylalanine.

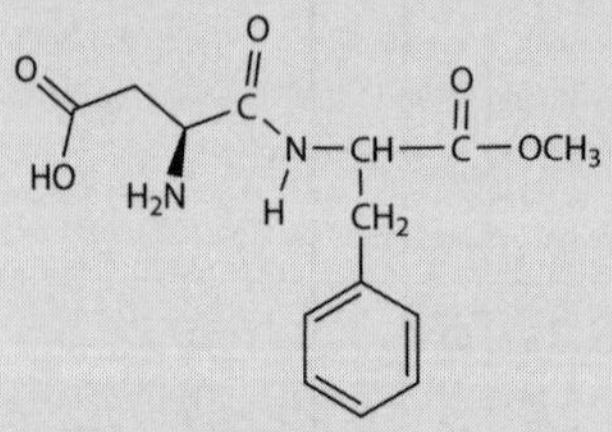

Figure 10.5 Aspartame.

KeyPoints

- Proteins and peptides are linear polymers of amino acids.
- There are 20 different types of amino acids that make up proteins.
- The amino acids in peptides and proteins are linked by *trans* peptide bonds.

- The unreacted amino group in the first amino acid *residue* in a peptide is referred to as the *N-terminus* of the peptide, and the unreacted carboxyl group in the last amino acid residue is referred to as the *C-terminus* of the peptide. The N-C_α-C(O)-N- C_α-C(O) ... spine of a peptide is referred to as the peptide *main chain*.
- The peptide bonds in peptides and proteins have substantial double-bond character and so can adopt either *cis* or *trans* configurations (Figure 10.4).
- Almost all the peptide bonds in peptides and proteins adopt the *trans* configuration; the *cis* configuration is generally only found for the peptide bonds involving proline.
- Peptides built from two amino acids are known as *dipeptides*; those built from three amino acids are *tripeptides*, and from four amino acids are tetrapeptides, and so on. Peptides involving a small number of amino acids are generally known as *oligopeptides*, and those involving many amino acids are known as *polypeptides*. The linked amino acid units within a peptide are known as *residues*.

Proteins

- Proteins are high-molecular-weight polymers of amino acids, typically involving one or more polypeptide chains of 50–500 amino acids.

■ Proteins generally have highly complex three-dimensional structures and the description of their structures is thus dealt with in a hierarchical way.

Protein primary structure

■ The lowest level in the *hierarchy of protein structure* is the *primary structure*. This is defined simply as the covalent structure of the protein, and it is conventionally presented as an ordered list of the protein's component amino acids, written from N-terminus to C-terminus – and including the details of any disulfide bonds formed between cysteine residues (Figure 10.6). The primary structure of a peptide or protein is sometimes referred to as its (amino acid) *sequence.*

Gly.Leu.Pro.Cys.Asn.Gln.Ile.Tyr.Cys

Figure 10.6
Primary structure of the nine-residue peptide hormone, oxytocin (used to induce labour in cases of difficult births). The N-terminal residue is glycine, the C-terminal residue, cysteine, and there is a disulfide bond bridging the side chains of cysteine 4 and cysteine 9.

■ The pattern of disulfide bonds formed in peptides and proteins that have multiple disulfide bonds is known as the *disulfide topology.*

Protein secondary structures

■ The second level in the hierarchy of protein structure is *secondary structure*. This is defined as the local folding of the polypeptide main chain stabilised by hydrogen bonding between its peptide bond NH and CO groups.
■ There are broadly three types of protein secondary structures: *helices, sheets* and *turns.*
■ There are different types of protein helices, each characterised by a particular pattern of main-chain hydrogen bonding, and the number of amino acid residues they have per turn.
■ The most common type of helix in proteins is the α-helix, which has 3.65 residues per turn, and hydrogen bonds between the peptide CO groups of residues (i) and the peptide NH groups of residues (i + 4) (Figure 10.7).
■ Helices can generally be wound in a clockwise or an anticlockwise sense; the former are referred to as *right-handed helices*, and the latter as *left-handed helices*. The α-helices found in proteins are always right-handed (Figure 10.8a); and this is a consequence of the stereochemistry of their L-amino acids.
■ Protein helices that contain a proline residue will tend to kink (Figure 10.8b). This is in part because proline is an imino acid

Figure 10.7

Pattern of main-chain hydrogen bonding in a protein α-helix. There are 3.65 residues per turn, and 13 atoms involved in each hydrogen bond ring (each hydrogen bond involving the peptide CO group of residue (i) and the peptide NH group of residue (i + 4)).

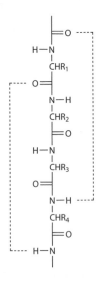

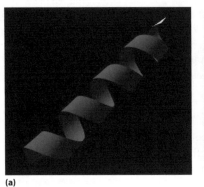 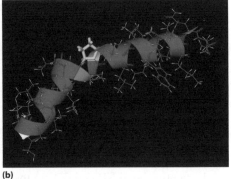

(a)   (b)

Figure 10.8 (a) Cartoon representation of a right-handed α-helix (the helix wound in a clockwise sense); (b) an α-helix kinked as a consequence of a proline residue (highlighted) in the fourth turn.

and has no NH group to participate in the main-chain hydrogen bonding, but also because of its bulky pyrrolidine side chain which, in an unkinked structure, would approach too close to atoms in the preceding turn of the helix.

■ The sheet secondary structures (more fully referred to as β-*pleated sheets*) can be *parallel* or *antiparallel*, as distinguished by the N- to C-terminal sense of their hydrogen-bonded chains.

■ Antiparallel β-sheets have their hydrogen-bonded chains (also referred to as β-*strands*) arranged with opposite sense, while parallel β-sheets have their β-strands arranged with the same sense (Figure 10.9).

■ The β-strands of a β-sheet are often arranged as the staves of a barrel (Figure 10.10).

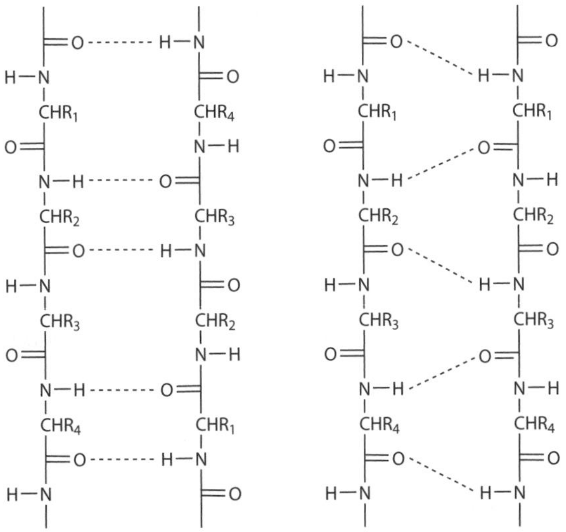

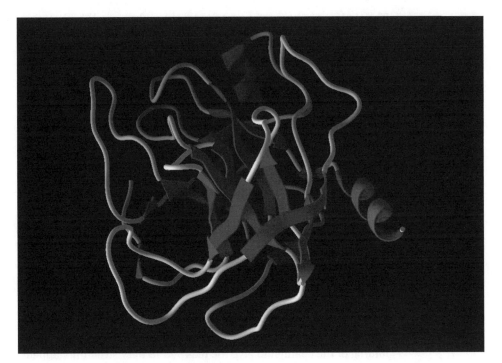

Figure 10.9
Antiparallel (left) and parallel (right)
β-pleated sheet structures. Hydrogen
bonds are shown as dashed lines.

Figure 10.10 Cartoon representation of a protein showing how the strands of a β-sheet (shown as arrows) are arranged in antiparallel fashion to form a barrel structure.

■ *Reverse turns* are the third major type of protein secondary structure, and these involve just four amino acid residues, stabilised by a single hydrogen bond formed between the peptide CO group of residue (1) and the peptide NH group of residue (1 + 3) (Figure 10.11).

Figure 10.11

Reverse turn involving four amino acid residues, with a hydrogen bond between the CO group of residue 1 and the NH group of residue 4

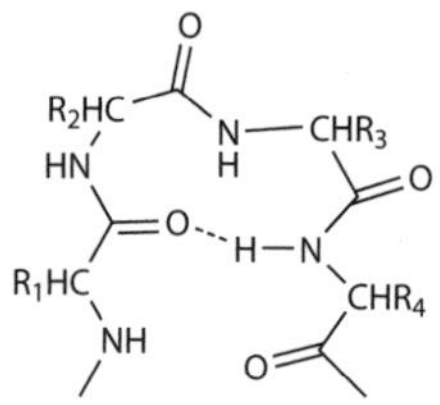

- Reverse turns cause a reversal in polypeptide chain direction. The residues at the second and third positions in the turn are frequently glycine or proline residues (the former because it has only H as a side chain and is conformationally highly flexible, and the latter because its pyrrolidine ring covalently dictates a 90° turn).

Protein tertiary structure

The tertiary structure of a protein describes the overall fold of its polypeptide chain(s) and is stabilised by non-covalent interactions between the residue side chains (viz. ionic interactions between residues with cationic and anionic side chains, hydrogen bonding between residues with polar side chains and van der Waals interactions between any residues in close contact in the protein interior).

In the folded structures of soluble proteins, those residues with polar side chains will tend to come to lie on the surface of the molecule (so as to allow their favourable interaction with the surrounding solvent water), while the residues with non-polar side chains will tend towards the interior of the structure – then being shielded from the solvent water. The removal of the non-polar residues to form this so-called *hydrophobic core* is the major driving force in protein folding. The removal of the non-polar residues from contact with water prevents these groups from interacting unfavourably with the surrounding solvent (decreasing its entropy), and thereby allows the liberated solvent greater freedom (the entropy of the solvent thus increasing).

In most cases, protein folding is *spontaneous* – that is, the cell acts only to assemble the protein polypeptide chain(s) and the folding of the chain(s) to give the required tertiary structure happens automatically, being dictated only by the sequence of the amino acids in the chain(s).

In large proteins (involving 100 residues or more) it is commonly found that the polypeptide chain(s) fold to form two or more *structural domains* (Figure 10.12). These domains are compact,

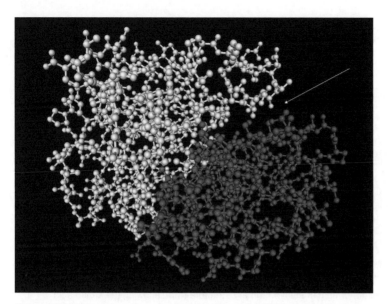

Figure 10.12 Structure of trypsin. A molecular model of the proteolytic enzyme trypsin – the N-terminal and C-terminal halves of the polypeptide chain are coloured white and grey, respectively. The active site of the enzyme, located in the cleft between the two domains, is indicated by an arrow.

semi-independent structural units, which very often fold autonomously. In many cases each has its own hydrophobic core.

Such domains often form functional units in proteins, associated, for example, with the binding of nucleotides like ATP, FAD or NADP, or in the binding of larger entities like oligosaccharide chains or DNA.

The hierarchy of structure for the hormone insulin is shown in Figure 10.13, and the basic domain structure of an antibody molecule, immunoglobulin G, is shown in Figure 10.14.

Protein quaternary structure

Many proteins exist as assemblies of polypeptide chains and, when ordered within the larger assembly, these folded polypeptides are referred to as *protein subunits*. The assemblies of subunits are known as *oligomers*, and the arrangement of the protein subunits within an oligomer is referred to as the protein's *quaternary structure*. The arrangement of subunits within a protein oligomer is maintained by virtue of non-covalent (ionic, hydrogen bond and van der Waals) interactions between the protein side chains.

A-chain

Gly.Ile.Val.Glu.Gln.Cys.Cys.Thr.Ser.Ile.Cys.Ser.Leu.Tyr.Gln.Leu.Glu.Asn.Tyr.Cys.Asn

Phe.Val.Asn.Gln.His.Leu.Cys.Gly.Ser.His.Leu.Val.Glu.Ala.Leu.Tyr.Leu.Val.Cys.Gly.Glu.Arg.Gly.Phe.Phe.Tyr.Thr.Pro.Lys.Thr

B-chain

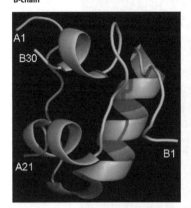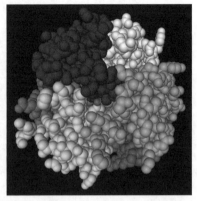

Figure 10.13 Structure of human insulin. Primary structure (top), showing the amino acid sequences of its two constituent polypeptide (A- and B-) chains, with the intra- and interchain disulfide bonds shown; cartoon (lower left) showing the secondary and tertiary structures of the molecule, with α-helices shown as coiled ribbons; space-filling model (lower right), showing the quaternary structure of the molecule, with the different subunits in the hexamer shown in various shades of white and grey.

Figure 10.14
Structure of an antibody, immunoglobulin G (IgG). The molecule is made up of four polypeptide chains: two light chains and two heavy chains. Each light chain is folded with two domains, one variable domain and the other constant domain. Each heavy chain has one variable domain and three constant domains. The four polypeptide chains are held together by disulfide bonds. The antigen-binding sites are found at the tips of the V_L and V_H domains.

disulfide bonds

C = constant domains

V = variable domains

L = light chains

H = heavy chains

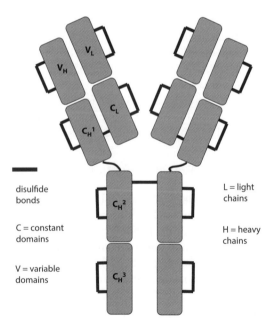

Some proteins have a quaternary structure that involves the assembly of a number of identical subunits. Such proteins include the hormone insulin, which is stored within the beta cells of the pancreas as *hexamers* – that is, assemblies of six identical insulin subunits.

Other proteins have quaternary structures that involve an assembly of different subunits. Such proteins include the oxygen transport protein, haemoglobin, which comprises two alpha subunits and two beta subunits, together arranged as a *tetramer*.

There are several possible advantages gained by proteins that have quaternary structure, including:

- an economy of DNA (since a given gene can be transcribed and translated multiple times to yield the required set of subunits)
- a reduced error in synthesis of the protein (because the cell makes fewer errors when transcribing shorter stretches of DNA)
- the possibility for *allosteric interactions* (wherein the binding of ligand(s) to the individual subunits in a protein oligomer can modify the ease with which subsequent ligand molecules bind to the other subunits)
- more efficient storage of the protein (since the protein assemblies can be packaged into a smaller space, and with a reduced level of solvent involved).

Proteins like the immunoglobulins that function within the blood do *not* have quaternary structure because their high dilution in the plasma would lead to rapid dissociation of their assembled subunits.

KeyPoints

- The primary structure of a protein is its covalent structure.
- The secondary structure of a protein describes its local folding as helices, beta sheets and reverse turns.
- The tertiary structure of a protein describes its overall three-dimensional structure.
- The quaternary structure of proteins describes the number and arrangement of their subunits as oligomers.

Protein denaturation

When a protein is subjected to some form of stress it is likely to *unfold* or *denature*, which means simply that it loses its secondary and/or tertiary and/or quaternary structure. Denatured proteins will tend to lose their biological activity.

Stresses that are likely to result in protein denaturation include exposure to extremes of pH or high temperature, exposure to organic solvents and exposure to *chaotropic agents* like urea and guanidinium chloride (compounds that disrupt the non-covalent interactions that help stabilise the folded protein structure).

Protein assays

Methods for the quantitative determination of the protein content of samples include the *BCA, modified Lowry* and *Coomassie/Bradford assays*. The former two methods depend upon the chelation of copper ions by the protein and the detection of reduced copper; the latter method depends upon the protein binding of dye molecules and monitoring of the resulting colour change.

Fibrous proteins

Polypeptide chains that have multiple repeats within their amino acid sequences will tend not to fold as globular proteins but instead form *fibrous proteins*. Such proteins are generally inert structural or storage proteins and are insoluble in water.

Fibrous proteins are used in the construction of connective tissue, bone, tendons and muscle fibres.

Examples of fibrous proteins include keratin (the protein found in hair and nails), elastin (the protein found in elastic tissue) and collagen (the protein found in cartilage).

Collagen is the most abundant protein found in mammals (accounting for around one-third of the total protein content). It is produced by fibroblast cells and exists as long fibrils. Each collagen polypeptide chain (referred to as an α-peptide) has a highly unusual amino acid sequence, involving multiple repeats of the patterns Gly-Pro-X and Gly-X-Hyp, where X is any amino acid, and Hyp is hydroxyproline (a post-translational modification of proline). Each collagen α-chain is individually wound to form a *left-handed helix*, and three such α-chains are then wound together to form a *right-handed triple helix* or *superhelix*. A number of superhelices are then wound together in a right-handed *super-super coil* to form a *collagen microfibril*.

KeyPoints

- Globular proteins fold spontaneously such that the amino acid residues with apolar side chains congregate towards the interior of the molecule, forming a hydrophobic core.
- Fibrous proteins have multiple repeats within their amino acid sequences and form long, fibre-like structures.
- Integral membrane proteins have amino acid sequences that render them compatible with insertion in the lipid bilayers of cell and organellar membranes.

Integral membrane proteins

Integral membrane proteins, that is, proteins that are embedded within the cell membrane or within the membrane of a cell organelle, have folded structures that differ from those of the soluble globular proteins, in that their outer – lipid bilayer-facing – surfaces are predominantly made up of residues with non-polar side chains.

Such proteins include: the transporter proteins (that function to regulate the cell's import and export of nutrients and metabolites), receptor proteins (that trigger changes inside cells in response to changing external levels of circulating hormones, for example), cell adhesion proteins (that are involved in cell–cell recognition), and membrane-bound enzymes (for example, the enzymes involved in oxidative phosphorylation). In the case of the transporters, such proteins may feature a central solvent-filled channel lined with hydrophilic residues, and in the case of the membrane receptor proteins, there may be extracellular domains that bind a specific neurotransmitter or hormone.

Self-assessment

1. At pH 7, the peptide hormone vasopressin, which has the primary structure shown below, will possess a net charge of:

Cys.Tyr.Phe.Gln.Asn.Cys.Pro.Arg.Gly

a. −1
b. 0
c. +1
d. +2

2. The antihypertensive drug captopril, shown below, is a modification of the dipeptide:

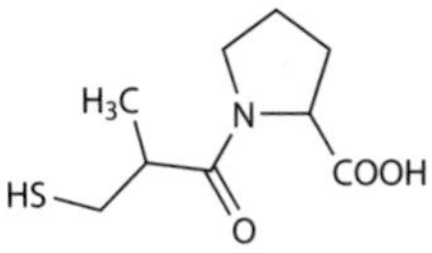

a. Pro.Ala
b. Ala.Pro
c. Pro.Cys
d. Ala.Phe

3. The diagram below shows the structure of:

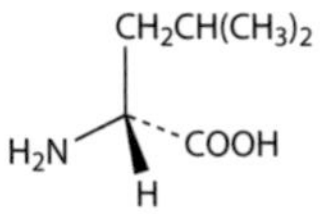

a. D-leucine
b. L-leucine
c. D-lysine
d. L-lysine

4. **Disulfide linkages in peptides and proteins are formed between the side chains of:**
a.   cystine residues
b.   cysteine residues
c.   serine residues
d.   methionine residues

5. **The endopeptidase enzyme EC 3.4.24.33 digests peptides by hydrolysing the peptide bonds on the N-terminal side of residues with anionic side chains. When this enzyme attacks delta sleep-inducing peptide, the primary structure of which is shown below,**

Trp.Ala.Gly.Asp.Ala.Ser.Gly.Glu

the products of digestion will be:
a.   Trp.Ala.Gly + Asp.Ala.Ser.Gly + Glu
b.   Trp.Ala.Gly.Asp + Ala.Ser.Gly.Glu
c.   Trp + Ala.Gly.Asp + Ala.Ser.Gly.Glu
d.   Trp.Ala.Gly + Asp + Ala.Ser.Gly + Glu

6. **The hormone insulin comprises two disulfide-linked polypeptide chains, each of which involves residues in α-helical conformation. From this statement we learn:**
a.   details that relate only to the peptide's primary structure
b.   details that relate to the peptide's primary and secondary structures
c.   details that relate only to the peptide's secondary structure
d.   details that relate to the peptide's secondary and tertiary structures

7. **The reverse turns in proteins commonly involve glycine because this residue:**
a.   is not optically active
b.   is conformationally very inflexible
c.   is an imino rather than an amino acid
d.   has only a hydrogen atom as a side chain

8. **In humans, vitamin C deficiency leads to an inability to hydroxylate proline to produce hydroxyproline, and this in turn leads to a reduced stability of the protein:**
a.   haemoglobin
b.   insulin
c.   keratin
d.   collagen

9. **The process of protein folding:**
a.   involves a gain in entropy of its constituent polypeptide chains
b.   occurs in such a way that the side chains of residues like lysine are removed from contact with the surrounding solvent
c.   is generally spontaneous
d.   leads to a decrease in entropy of the surrounding solvent

10. Integral membrane proteins:

a. are composed exclusively of residues with non-polar side chains

b. are found in the cell plasma membrane but not in the membranes of cell organelles

c. do not have any part of their structure exposed to the cell cytoplasm or to the extracellular fluid

d. include the proteins associated with cell–cell recognition

chapter 11
Carbohydrates and nucleic acids

Overview

After learning the material presented in this chapter you should:

- know the various systems of nomenclature relating to carbohydrates: be able to distinguish mono-, di-, oligo- and poly-saccharides; be able to distinguish aldose and ketose sugars; be able to distinguish trioses, tetroses, pentoses and hexoses; be able to distinguish pyranose and furanose sugars
- understand the stereochemistry of monosaccharides, be able to recognise the L- and D-forms of sugars, and their α- and β-anomers
- know how monosaccharides can be covalently linked through glycosidic bonds to form disaccharides, oligosaccharides and polysaccharides
- know the structures of the monosaccharides: glyceraldehyde, ribose, deoxyribose, glucose, galactose and fructose
- know the structures of the disaccharides: sucrose, lactose, maltose and cellobiose
- know the structures of the polysaccharides: starch (and its components amylose and amylopectin), glycogen, cellulose and heparin
- understand the difference between reducing and non-reducing sugars, and how Benedict's and Fehling's tests can be used to test for monosaccharides
- appreciate the general form of glycosaminoglycans and glycoproteins
- know the structures of the purines, adenine and guanine, and the pyrimidines, cytosine, thymine and uracil
- be able to distinguish, and know the structures of ribo- and deoxyribo-nucleosides and nucleotides
- know the Watson–Crick base pairings in DNA (and RNA) and appreciate the double-helix structure of DNA.

Carbohydrates

- Carbohydrates are otherwise known as *saccharides*. They are polyhydroxy aldehydes and ketones, and generally – but not always – have the empirical formula, $C_x(H_2O)_y$.
- The lower-molecular-weight members of the class – commonly referred to as *sugars* – include *monosaccharides* and *disaccharides*.
- Monosaccharides are the simplest carbohydrates, in the sense that they cannot be hydrolysed to smaller carbohydrates. Examples of relevance in pharmaceutical science include: the three-carbon sugar, *glyceraldehyde*, the five-carbon sugars, *ribose* and *deoxyribose* and the six-carbon sugars, *glucose*, *galactose*, *fructose* and *mannose*.

■ Sugars that have three, four, five and six carbons are respectively known as *trioses, tetroses, pentoses* and *hexoses*.

■ Sugars that have an aldehyde group (–CHO) are known as *aldoses*, while those that involve a ketone group (R$_2$C=O) are known as *ketoses*.

These two forms of nomenclature are sometimes combined, so that glyceraldehyde and glucose, for example, which are aldehydes, are respectively described as an *aldotriose* and an *aldohexose*, while fructose, which has a keto-group, is described as a *ketohexose*.

The simplest monosaccharide, glyceraldehyde, has an aldehyde group at C-1, and hydroxyl groups attached at C-2 and C-3. It exists in two enantiomeric forms, D-glyceraldehyde and L-glyceraldhyde – the former, in Fischer projection, having the 2-hydroxy group drawn to the right, and the latter having it drawn to the left (Figure 11.1).

Monosaccharide sugars that have five or more carbons, e.g. ribose, glucose and fructose, can exist as open-chain molecules or as heterocylic ring compounds, and they too have L- and D-forms. These enantiomers have opposite configurations about the carbon atom furthest from the aldehyde/ketone carbon (Figure 11.1).

The ring forms of the pentoses and hexoses are created through a reaction between one of the molecule's hydroxyl groups and the aldehyde or ketone group. Sugars that are cyclised to form five-membered rings are known as *furanoses*, while those cyclised to form six-membered rings are known as *pyranoses* (Figure 11.2).

KeyPoints

■ Carbohydrates are divided into monosaccharides (which are the simplest form of carbohydrate), disaccharides (which comprise two covalently linked monosaccharide units), oligosaccharides (which comprise several covalently linked monosaccharide units), and polysaccharides (which comprise many covalently linked monosaccharide units).

■ Monosaccharides may be generically named according to the numbers of carbons they contain (three: trioses, four, tetroses, etc) and/or according to whether they possess an aldehyde or ketone group (aldoses vs ketoses).

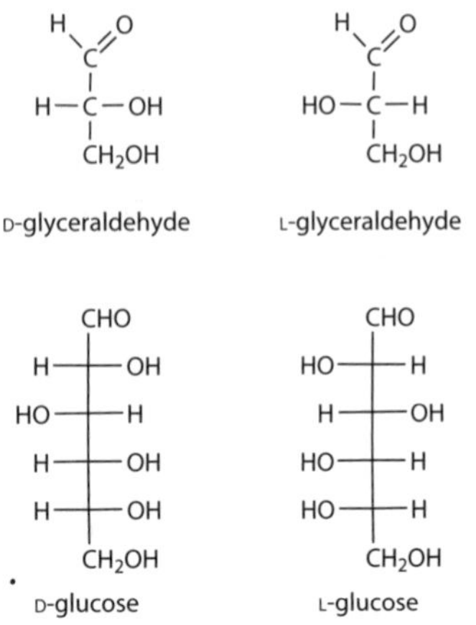

Figure 11.1
L- and D-forms of the aldotriose, glyceraldehyde, and the aldohexose, glucose.

D-glyceraldehyde

L-glyceraldehyde

D-glucose

L-glucose

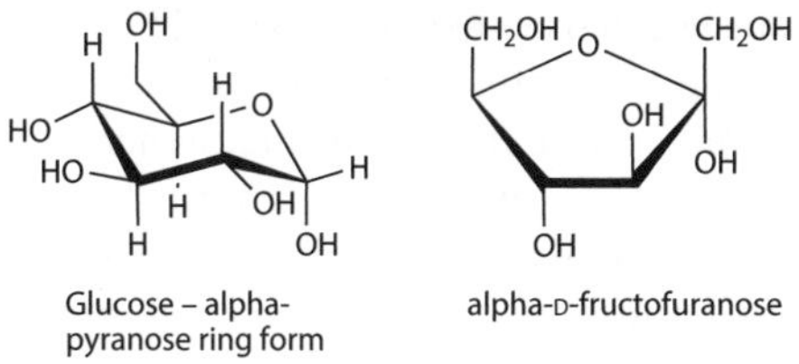

Glucose – alpha-pyranose ring form

alpha-D-fructofuranose

Figure 11.2
Pyranose form of glucose (left) and furanose form of fructose (right).

When the open-chain forms of sugars are cyclised to create the corresponding ring forms, the aldehyde/ketone carbon forms a new chiral centre. This carbon is referred to as the *anomeric carbon*, and the two possible stereoisomers formed on cyclisation are distinguished as the α-*anomer* and the β-*anomer* (Figure 11.3).

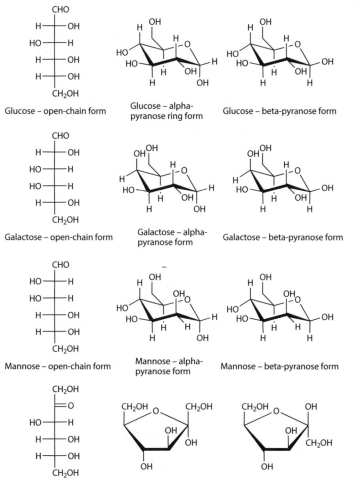

Glucose – open-chain form

Glucose – alpha-pyranose ring form

Glucose – beta-pyranose form

Galactose – open-chain form

Galactose – alpha-pyranose form

Galactose – beta-pyranose form

Mannose – open-chain form

Mannose – alpha-pyranose form

Mannose – beta-pyranose form

D-Fructose – open-chain form

alpha-D-fructofuranose

beta-D-fructofuranose

Figure 11.3
Common monosaccharides in their open-chain and various cyclised forms.

Tip

The D- and L- descriptors of sugars are *not* synonymous with the labels d- and l-. The latter, lower-case, labels – now obsolete – were once used to specify the direction in which a sugar molecule rotated the plane of plane-polarised light.

The sugar glucose, therefore, exists in an open-chain form, either as D-glucose or L-glucose, and each of these two enantiomers can cyclise to give a (pyranose) ring form, each of which can exist as α- and β-anomers (Figures 11.1 and 11.3). The glucose anomers also constitute different *epimers* of glucose – that is, they are stereoisomers that differ in configuration at a single chiral centre.

The covalent linkage of monosaccharide sugars to form *disaccharides* involves the formation of *glycosidic bonds*. If the glycosidic bond is formed through a condensation reaction involving the hydroxyl group on the C-4 carbon of one sugar and the hydroxyl group on the C-1 (anomeric) carbon of the second sugar, the linkage is referred to as a 1→4 linkage. Since the C-1 hydroxyl group may have the α- or β-orientation, the glycosidic linkage can thus be a β-(1→4) or an α-(1→4) linkage.

Glycosidic linkages may also be formed between the hydroxyls on the C-1 and C-6 positions, in which case the linkage is referred to as a 1→6 linkage.

Disaccharides of importance in pharmaceutical science include maltose (formed through the (1→4) linkage of two α-D-glucose molecules), lactose (formed through the (1→4) linkage of β-D-galactose and D-glucose), sucrose (formed through the 1→2 linkage of α-D-glucose and D-fructose) and cellobiose (formed through the (1→4) linkage of two β-D-glucose molecules) (Figure 11.4).

Any aldose sugar, and any ketose sugar that is able to isomerise to form an aldehyde group in solution, is known as a *reducing sugar*. The aldehyde group allows the compound to act as a reducing agent.

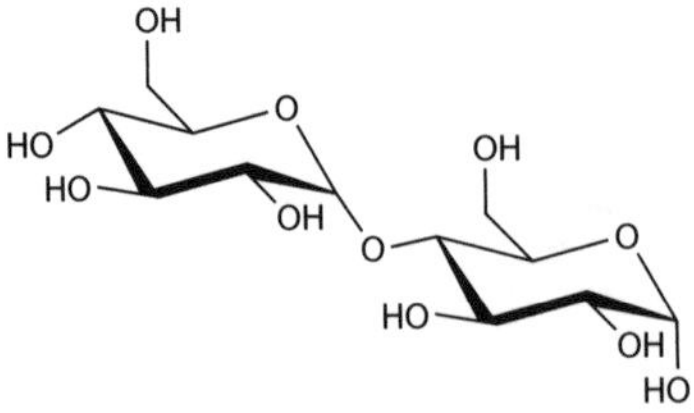

Maltose = α-D-glucopyranosyl-(1→4)-D-glucose

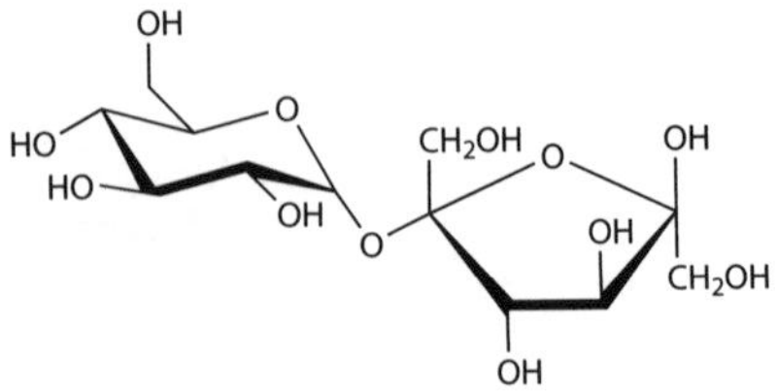

Sucrose = α-D-glucopyranosyl-(1→2)-D-fructose

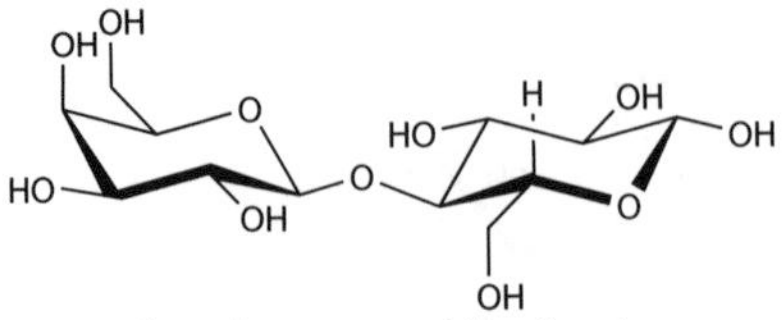

Lactose = β-D-galactopyranosyl-(1→4)-D-glucose

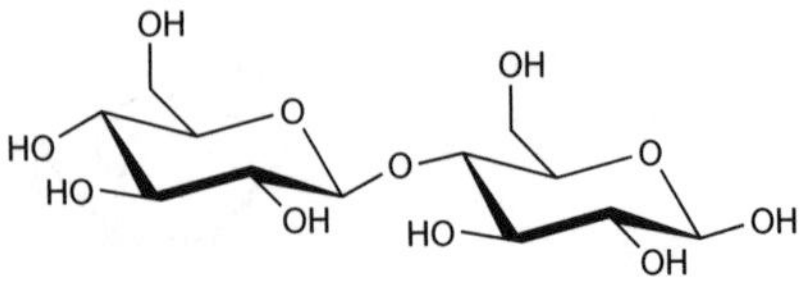

Cellobiose = β-D-glucopyranosyl-(1→4)-D-glucose

Figure 11.4 Disaccharides.

Monosaccharides that are reducing sugars include: glyceraldehyde, glucose and galactose.

Disaccharides that are reducing sugars include lactose and maltose. In these disaccharides, one of the constituent monosaccharides has an open-chain form bearing an aldehyde group.

Disaccharides that are non-reducing sugars include sucrose and trehalose. In these disaccharides the consitituent monosaccharides are glycosidically linked through their anomeric carbons and cannot open to form an aldehyde group.

Polysaccharides are high-molecular-weight carbohydrate polymers formed through the polymerisation of monosaccharides. If they are formed from just a single type of monosaccharide, they are known as *homopolysaccharides*, and if they involve a combination of two or more monosaccharides they are known as *heteropolysaccharides*.

Homopolysaccharides of importance in pharmaceutical science include: *cellulose*, which is the principal component of plant cell walls; *glycogen*, which is the principal storage polysaccharide in animals; and *amylose* and *amylopectin*, which together make up the plant storage polysaccharide, *starch* (Figure 11.5).

Glycosaminoglycans (also referred to as *mucopolysaccharides*) are high-molecular-weight, linear polysaccharides, based on repeating disaccharide units involving an amino sugar and either galactose or a uronic acid. The amino sugars involved are most often *N-acetylglucosamine* (GlcNac) or *N-acetylgalactosamine* (GalNac); the uronic acids involved are most often *glucuronic acid* (GlcUA: Figure 11.6) or iduronic acid (IdoUA).

The glycosaminoglycan, *hyaluronic acid,* which is a major constituent of synovial fluid, is based on the repeating disaccharide unit: →4) GlcAβ(1→3)GlcNAcβ(1→.

The glycosaminoglycan, *heparin,* which is used clinically as an anticoagulant, and *in vivo* functions to aid histamine storage in mast cells, is based on a repeating disaccharide unit involving sulfated derivatives of IdoUA and glucose (Figure 11.6).

Proteins that have oligosaccharide chains attached to their polypeptide chains are known as *glycoproteins*.

Tip

Reducing sugars can be detected/visualised by reacting with *Benedict's reagent* or *Fehling's reagent*. These reagents contain a copper(II) complex which appears blue in colour, and when heated with a reducing sugar they form a brick red precipitate of copper(I) oxide. Benedict's reagent comprises a single solution, while Fehling's reagent is prepared from two solutions that are mixed in equal proportion immediately prior to use. These tests can be used to demonstrate the presence of glucose in urine, thereby providing a test for diabetes.

KeyPoints

- Monosaccharides are covalently linked through glycosidic bonds to form disaccharides, oligosaccharides and polysaccharides.
- Polysaccharides based on a single monosaccharide monomer are homopolysaccharides; those based on two or more monosaccharide monomers are heteropolysaccharides.
- Glycosaminoglycans are heteropolysaccharides and are most often based on repeating disaccharides formed from an amino sugar and a uronic acid.
- Glycoproteins are proteins that have oligosaccharide chains attached to the side chains of serine and/or threonine and/or asparagine residues.

Figure 11.5 Polysaccharides.

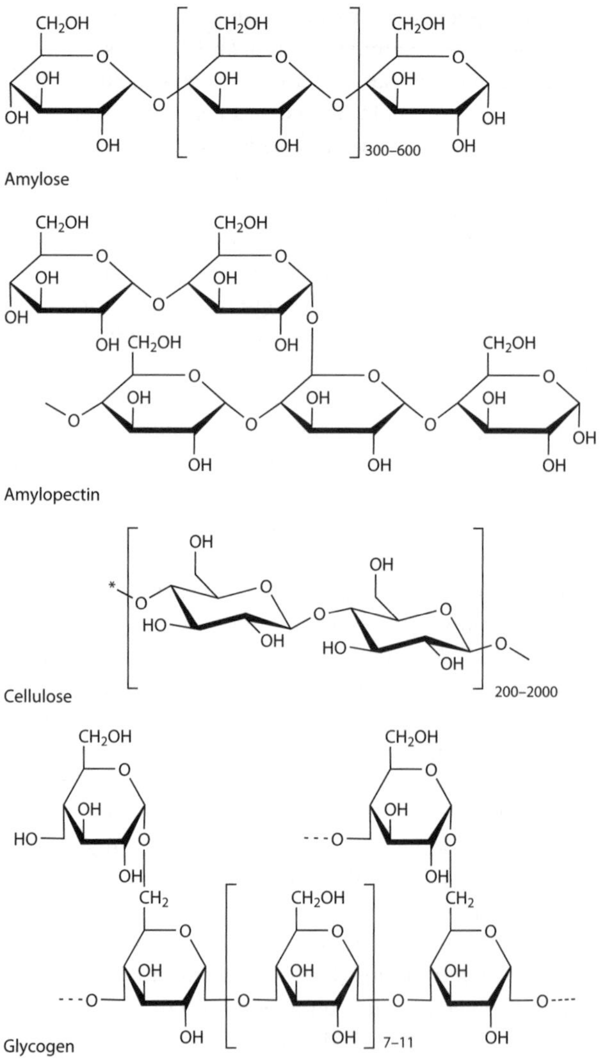

The oligosaccharide chains are attached to the polypeptide chains in glycoproteins either via the side chain amide group of an asparagine residue (*N-glycosylation*) or *via* the side chain hydroxyl group of a serine or threonine residue (*O-glycosylation*).

Many integral membrane proteins are glycoproteins, and these proteins have their oligosaccharide chains attached on their extracellular domains. Such proteins are involved in cell–cell interactions/recognition.

The oligosaccharide chains attached to human glycoproteins generally involve *sialic acid* (otherwise known as *N-acetyl-neuraminic acid* (Figure 11.6), and it is to this sugar that the influenza virus binds, and which the virus exploits to gain entry into human cells.

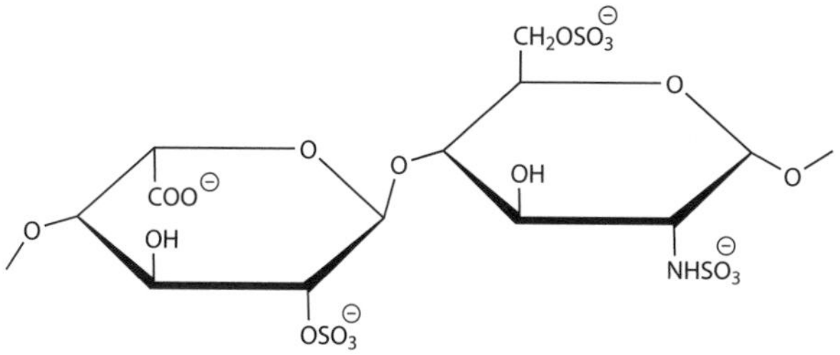

Figure 11.6
Sugars involved in glycosaminoglycans and in the oligosaccharide components of glycoproteins.

Major disaccharide unit in heparin = 2-O-sulfo-α-L-iduronic acid linked to 2-deoxy-2-sulfamido-α-D-glucopyranosyl-6-O-sulfate

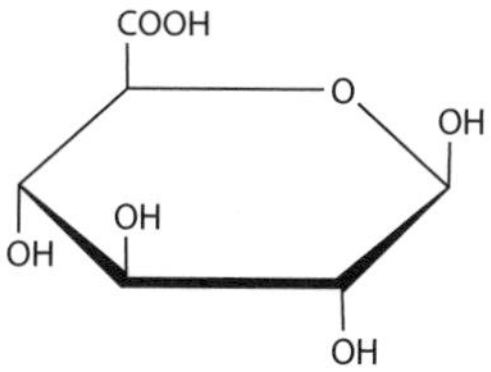

β-D-glucuronic acid 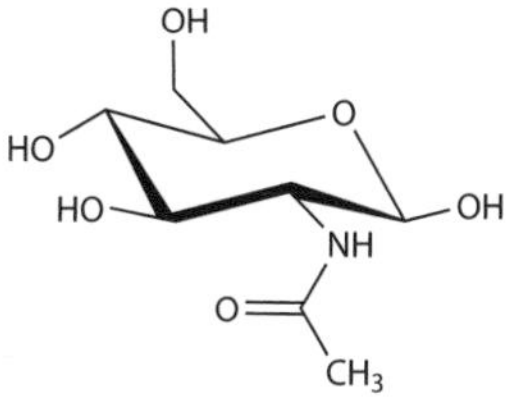   N-acetyl-glucosamine

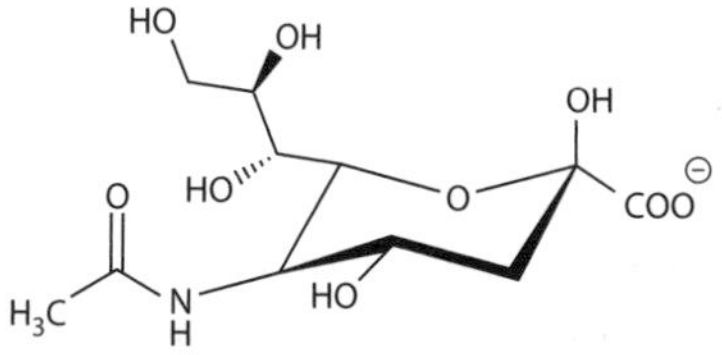

N-acetyl-neuraminic acid (sialic acid)

Nucleic acids

Nucleic acids are polymeric molecules, constructed from monomers known as *nucleotides*. They are the molecules in living organisms that encode, transmit and express genetic information. The two principal forms of nucleic acid are deoxyribonucleic acid (DNA) and ribonucleic acid (RNA).

Nucleic acids (in particular, DNA) provide the intracellular target for numerous anticancer, antiviral and antibacterial drugs, and so an understanding of this class of biomolecule is essential in many areas of pharmaceutical science.

The nucleotide building blocks of nucleic acids have three molecular components: a nitrogenous heterocyclic base (either a *purine* or a *pyrimidine*: Figure 11.7), covalently bonded to the C-1 hydroxyl group of a pentose sugar (either *ribose* or *2′-deoxyribose*: Figure 11.8), and a *phosphate* group attached at the 5′ hydroxyl group of the sugar.

The pyrimidines involved include: *cytosine, thymine* and *uracil*; the purines involved include: *adenine* and *guanine* (Figure 11.7).

The pyrimidines found in DNA include cystosine and thymine, whereas in RNA they include cytosine and uracil.

Figure 11.7
Purine (adenine, guanine) and
pyrimidine (cytosine, thymine, uracil)
bases involved in DNA and RNA.

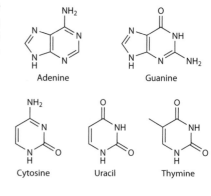

Adenine   Guanine

Cytosine   Uracil   Thymine

Figure 11.8
Nucleic acid sugar
components.

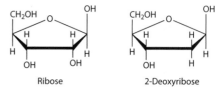

Ribose   2-Deoxyribose

Tip

There are a number of antiviral drugs that are nucleoside analogues, e.g. *acyclovir*, which is used to treat herpes simplex infections, and *zidovudine*, also referred to as *AZT*, which is used to delay the development of AIDS following HIV infection.

Nucleosides are compounds closely related to nucleotides; they involve one of the purine or pyrimidine heterocyclic bases, covalently linked to ribose or deoxyribose. They thus differ from nucleotides in not having a phosphate group attached at the 5′ position of the sugar (Figure 11.9).

DNA and RNA are composed of deoxyribonucleotides or ribonucleotides, respectively, joined together through *phosphodiester bonds* (Figure 11.10).

The phosphate groups in DNA and RNA account for their acidic properties, and the negative charges on these groups make the molecules polyanions.

In DNA, the polynucleotide chains are wound together as a *double helix* (Figure 11.11), with one chain running in the 5′ to 3′

Figure 11.9
Nucelotide and
nucleoside structure.

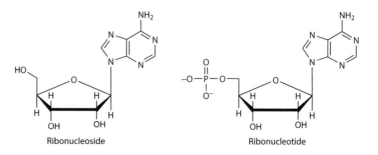

Ribonucleoside   Ribonucleotide

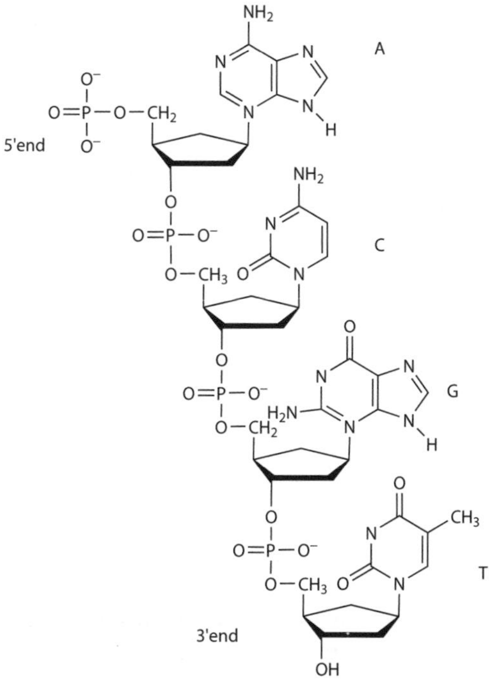

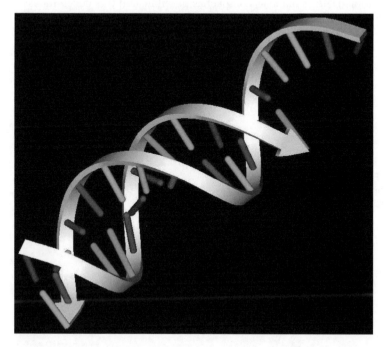

Figure 11.10
Polynucleotide structure showing the phosphodiester backbone, and indicating the chain sense (5′ to 3′).

KeyPoints

- DNA and RNA are polyanionic polymers based on monomers known as nucleotides.
- Nucleotides comprise a phosphate group, either ribose or deoxyribose sugar, and a purine or pyrimidine base.
- DNA comprises two antiparallel polynucleotide chains, entwined to form a double helix, and stabilised by hydrogen bonds between opposing bases, with the pairing adenine:thymine, guanine:cytosine.

Figure 11.11 Schematic of DNA double helix (arrows indicating the chain sense of the entwined polynucleotide chains; the hydrogen-bonded base pairs are shown as the rungs within this coiled ladder).

direction, and the other arranged in antiparallel fashion, running in the 3′ to 5′ direction.

It is the order of purine and pyrimidine bases along the polynucleotide strands of DNA that constitutes the *genetic code*.

The particular polynucleotide strand whose *base sequence* is transcribed in a *gene* to produce a particular protein is known as the *sense strand*; and the other strand is known as the *antisense strand*.

The base sequence of the sense strand in DNA is transcribed as messenger or mRNA, and from this, on the ribosome – which is composed of ribosomal or rRNA – a protein polypeptide chain is synthesised via the amino acids attached to transfer or tRNA molecules.

The two polynucleotide chains in DNA are held together via hydrogen bonds formed between complementary bases. These base pairs are very specific; they involve adenine and thymine, and guanine and cytosine (in DNA) or guanine and uracil (in RNA).

The pattern of hydrogen bonds involved in the standard (Watson–Crick) base pairings is shown in Figure 11.12.

The various forms of RNA generally exist as single-stranded molecules, with their three-dimensional structures stabilised by base pair hydrogen bonding within the same polynucleotide chain.

The efficient packaging of DNA within a cell's nucleus demands that the DNA is made as compact as possible, and this necessitates that the double helix is *supercoiled* (Figure 11.13).

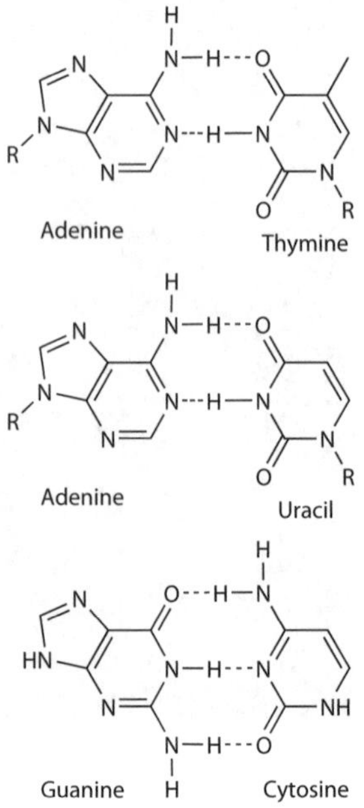

Figure 11.12
Watson–Crick base pairings of nucleic acid bases.

Adenine
Thymine

Figure 11.13
A schematic of supercoiled (circular) DNA.

Adenine
Uracil

Guanine   Cytosine

Self-assessment

1. Which of the following will not give a positive result in Fehling's test?
a. lactose
b. sucrose
c. glucose
d. fructose

2. Which of the following is not a homopolymer of glucose?
a. cellulose
b. glycogen
c. heparin
d. starch

3. The anticancer drug Efudex (shown below) is a derivative of:

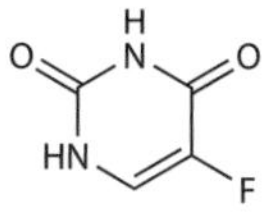

a. adenine
b. thymine
c. uracil
d. guanine

4. Which of the following is not a Watson–Crick base pair?
a. guanine–cytosine
b. adenine–thymine
c. adenine–uracil
d. uracil–cytosine

5. Which of the following statements is *not* correct?
a. α-D-glucose and β-D-glucose are enantiomers
b. α-D-glucose and β-D-glucose are epimers
c. α-D-glucose and β-D-glucose are anomers
d. α-D-glucose and β-D-glucose are diastereomers

6. The antiretroviral drug zidovudine (shown below) is:

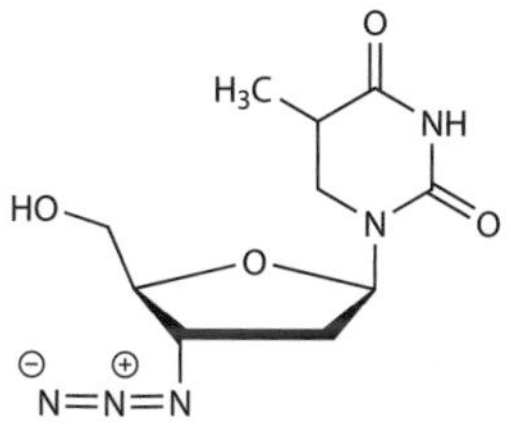

a. a purine nucleoside analogue
b. a purine nucleotide analogue

c.   a pyrimidine nucleoside analogue
d.   a pyrimidine nucleotide analogue

7.   **Which of the following is not a hexose sugar?**
a.   ribose
b.   mannose
c.   glucose
d.   galactose

8.   **Which of the following is not a pyrimidine?**
a.   adenine
b.   thymine
c.   cytosine
d.   uracil

9.   **Unlike DNA, RNA:**
a.   does not involve phosphodiester bonds
b.   carries a net positive charge at pH 7
c.   contains thymine rather than uracil
d.   involves the pentose sugar, ribose

10. **Lactose is:**
a.   β-D-galactosyl-(1→4)-β-D-glucose
b.   β-D-glucosyl-(1→4)-β-D-galactose
c.   β-D-galactosyl-(1→4)-α-D-glucose
d.   β-D-glucosyl-(1→4)-α-D-galactose

chapter 12
Lipids and steroids

Overview

After learning the material presented in this chapter you should:

- know the various compounds that make up the class of lipids
- know the various roles played by lipids in living systems
- know the structures of the common simple lipids and their physical properties
- know the structures and understand the physical properties of the common membrane lipids: glycerophospholipids and sphingolipids
- understand and be able to predict the aggregates likely to be formed in aqueous media by lipid, based on a consideration of their molecular shape
- know the ring-labelling and atom-numbering schemes applied to steroids
- be able to interpret two-dimensional representations of steroids to recognise alpha- and beta-substituents
- be able to recognise the class to which a steroid belongs (pregnane, cholestane, androstane, cholane or estrane)
- be able to interpret and construct the IUPAC names of steroids
- know the structures and properties of the cardiac glycosides.

Lipids

Lipids comprise all compounds that can be solubilised from biological cells/tissues by treatment with organic solvents (such as chloroform, methanol, hexane). They include *fats*, *waxes*, *sterols*, some vitamins (such as *vitamin A*, *vitamin D*, *vitamin E* and *vitamin K*), *glycerides*, *glycerophospholipids* and *sphingolipids*.

Lipids serve a variety of different functions in living systems:

- they provide *thermal insulation* – as a major constituent of *adipocytes* in *adipose tissue*
- they provide *electrical insulation* – as a major constituent of the *Schwann cells* that make up the *myelin sheath* surrounding the axons of nerve cells
- they provide for *cell–cell recognition* – the carbohydrate moieties of *glycolipids* presenting as *antigens* on the external surfaces of cell membranes
- they provide a *physical barrier*, as structural components of *cell membranes*, separating the external environment and the cytoplasm of cells – the core of the cell membrane comprising a *lipid bilayer*
- they provide a cellular energy reserve.

Simple lipids

Simple lipids include saturated, monounsaturated and polyunsaturated fatty acids and mono-, di- and tri-glycerides.

The common *saturated fatty acids* found in mammalian tissues include *myristic acid, palmitic acid* and *stearic acid* (Table 12.1).

The common *unsaturated fatty acids* found in mammalian tissues include *palmitoleic acid*, and *oleic acid* – these fatty acids have just one double bond – and *linoleic acid* and *linolenic acid*, which have two and three double bonds, respectively (Table 12.2).

Linoleic acid is an example of an *omega-6 fatty acid* (the first carbon atom in its double bond is the sixth carbon atom along from the methyl terminus/omega carbon end of the hydrocarbon chain: Table 12.2).

Linolenic acid is an example of an *omega-3 fatty acid* (the first carbon atom in its double bond is the third carbon atom along from the methyl terminus/omega carbon end of the hydrocarbon chain: Table 12.2).

The double bonds involved in mammalian unsaturated fatty acids always have the *cis* (never the *trans*) configuration.

Regardless of the number of double bonds involved in a mammalian unsaturated fatty acid, the first is always located between carbons 9 and 10; the second (if present) involves carbons 12 and 13; the third (if present) involves carbons 15 and 16, and so on.

The double bonds in a polyunsaturated fatty acid are thus never conjugated. Such an arrangement would make the compounds much too chemically reactive.

Table 12.1 Examples of saturated fatty acids

Fatty acid	Chemical structure	Shorthand code*	Melting point (°C)
Myristic acid	$CH_3(CH_2)_{12}COOH$	$C_{14:0}$	52.0
Palmitic acid	$CH_3(CH_2)_{14}COOH$	$C_{16:0}$	63.1
Stearic acid	$CH_3(CH_2)_{16}COOH$	$C_{18:0}$	69.6

* In the shorthand codes, the number before the colon indicates the carbon chain length, wheras the number after the colon indicates the number of double bonds in the chain.

Table 12.2 Examples of unsaturated fatty acids

Fatty acid	Chemical structure	Shorthand code*	Melting point (°C)
Palmitoleic acid	$CH_3(CH_2)_5CH=CH(CH_2)_7COOH$	$\Delta^9 C_{16:1}$	−0.5
Oleic acid	$CH_3(CH_2)_7CH=CH(CH_2)_7COOH$	$\Delta^{9,12} C_{18:1}$	13.4
Linoleic acid	$CH_3(CH_2)_4(CH=CHCH_2)_2(CH_2)_6COOH$	$\Delta^{9,12,15} C_{18:3}$	−9.0
Linolenic acid	$CH_3CH_2(CH=CHCH_2)_3(CH_2)_6COOH$	$\Delta^{9,12,15} C_{18:3}$	−17.0

* In the shorthand codes, the number before the colon indicates the carbon chain length, whereas the number after the colon indicates the number of double bonds in the chain. The Greek symbol delta (Δ) indicates an unsaturated fatty acid, with the superscript numbers indicating the first carbon atom involved in each of the double bonds.

The dietary ratio of omega-6 : omega-3 fatty acids has implications for health, with high ratios generally being bad.

The melting point of a fatty acid varies with the hydrocarbon chain length and/or the degree of unsaturation/number of double bonds in the chain:

- The melting point increases with increasing hydrocarbon chain length (longer chains interact more extensively via van der Waals interactions and so require greater energy input for chain melting).
- The melting point decreases with increasing degree of chain unsaturation (double bonds disrupt the packing of the hydrocarbon chains, reducing van der Waals interactions and thus decreasing the energy input required for chain melting).

The shorthand codes and melting points of the fatty acids most common in humans are given in Tables 12.1 and 12.2.

Lower organisms like bacteria and fungi very often possess *hydroxylated fatty acids* and/or *branched fatty acids* (Table 12.3). Bacterial fatty acids are rarely polyunsaturated.

Glycerides are fatty acid esters of the *trihydric alcohol, glycerol* (Figure 12.1).

The melting point of a glyceride varies according to the nature of its component fatty acid(s). Glycerides that are esters of long-chain fatty acids will have higher melting points than those that involve

Table 12.3 Examples of bacterial fatty acids

Fatty acid	Chemical structure
13-methyltetradecanoic acid	$(CH_3)_2CH(CH_2)_{11}COOH$
β-hydroxymyristic acid	$CH_3(CH_2)_{10}CH(OH)CH_2COOH$

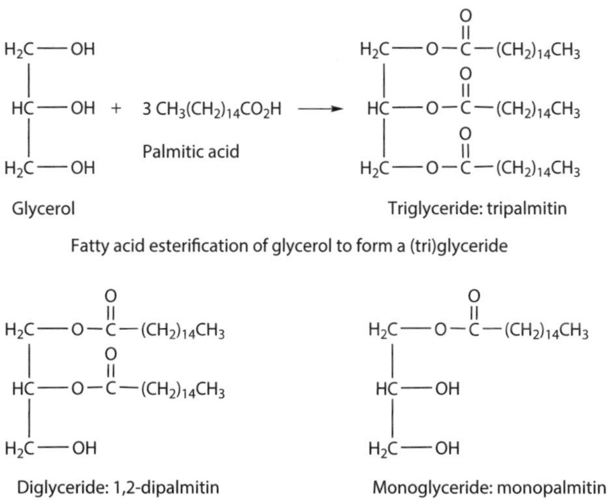

Figure 12.1 Glycerides.

Fatty acid esterification of glycerol to form a (tri)glyceride short-chain fatty acids, and saturated fatty acid glycerides will have higher melting points than unsaturated fatty acids. Thus, tristearin, which has stearic acid attached at each glycerol OH, has a melting point of 54°C, while triolein, which has oleic acid attached at the three OH groups, has a melting point of −32°C.

Triglycerides have fatty acids ester-linked at all three glycerol hydroxyls. *Diglycerides* have only two of the glycerol hydroxyls esterified and *monoglycerides* have just one of the glycerol hydroxyls esterified (Figure 12.1).

KeyPoint

- Simple lipid melting points vary according to the length and degree of unsaturation of their fatty acid hydrocarbon chains: greater unsaturation and shorter chain lengths result in lower melting points.

Simple triglycerides have the same type of fatty acid attached at each of the glycerol hydroxyls. *Mixed triglycerides* have different types of fatty acid attached at the glycerol hydroxyls.

Fats involve mixtures of simple lipids that are solid at room temperature, e.g. cocoa butter. *Oils* are mixtures of simple lipids that are liquid at room temperature, e.g. olive oil and cod liver oil. *Waxes* are mixtures of simple lipids that are semisolid at room temperature, e.g. beeswax.

Membrane lipids

The lipids found in cell membranes include:
- *glycerophospholipids* (fatty acid esters of *phosphatidylglycerol*), e.g. *phosphatidyl-cholines, phosphatidyl-serines, phosphatidyl-ethanolamines* (Figure 12.2)
- *sphingolipids* (fatty acid esters of *sphingosine*), e.g. *sphingomyelins, ceramides, cerebrosides* and *gangliosides* (Figure 12.3).

Phosphatidylcholines – also referred to as *lecithins* – are widely used as pharmaceutical excipients, and can be obtained from egg (*egg lecithin*) and soy (*soy lecithin*).

All classes of membrane lipids are *amphiphilic* compounds, containing both a *hydrophobe* moiety (comprising their hydrocarbon chains) and some form of hydrophilic/polar *head group*.

Ceramides are *N*-acyl fatty acid derivatives of sphingosine.

Figure 12.2
Distearoyl glycerophospholipids.

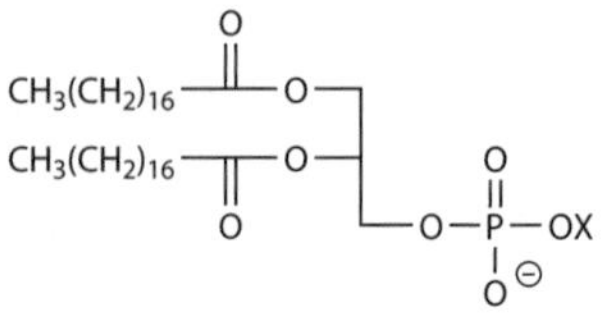

	X
Phosphatidic acids	-H
Phosphatidylethanolamines	$-(CH_2)_2NH_3^+$
Phosphatidylcholines (lecithins)	$-(CH_2)_2N(CH_3)_3^+$
Phosphatidylglycerols	$-CH_2CH(NH_3^+)COO^-$
Cardiolipins	-phosphatidylglycerol

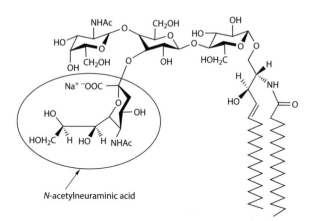

Figure 12.3
Sphingolipids and glycolipids.

Sphingosine

Sphingomyelin (sphingophospholipid)

Ceramide

Galactose

Galactocerebroside (a ceramide with galactose as polar head group)

N-acetylneuraminic acid

Ganglioside (GM2) (a ceramide with an oligosaccharide polar head group involving N-acetylneuraminic acid (otherwise known as sialic acid)

Sphingomyelins are ceramides that have a phosphatidyl head group; they are major components of the nerve cell myelin sheath.

Cerebrosides are the simplest form of *glycolipid*. They are ceramides that have a single sugar molecule attached as the polar head group. They include the galactocerebrosides, which have a galactose head group, and are found in high levels in nerve cell membranes.

Gangliosides are complex glycolipids; they are ceramides that have oligosaccharides as their polar head group. These glycolipids make up a significant fraction of brain lipid, and are also involved in cell–cell recognition. They include the antigens that are found on the

KeyPoint

- The fluidity and permeability of a cell membrane are determined by its lipid composition. That is, the relative proportions of lipids with long hydrocarbon (h/c) chains vs short h/c chains, the proportions of lipids with saturated vs unsaturated h/c chains, and the amount of sterol in the membrane.

surfaces of red blood cells – the antigens that give rise to our blood groups.

One or more of the sugar residues in the ganglioside head group will be *N-acetylneuraminic acid* (also known as *sialic acid*). Natural and semisynthetic gangliosides are currently considered as possible therapeutics for the treatment of neurodegenerative disorders.

The *influenza virus* exploits certain cell surface gangliosides as a means of gaining entry to cells and infecting them.

Lipid aggregates

In solution, amphiphilic lipids will tend to self-associate to form aggregates. The nature of the aggregates formed will vary according to the nature of the solvent and according to the shapes of the lipid molecules. The simplest types of aggregates formed include spherical *micelles, reverse micelles* and *monolayers* and *bilayers* (Figure 12.4).

In aqueous media, lipids that have a conical shape, i.e. that have a polar head group with a cross-sectional area that is larger than the cross-sectional area of their hydrophobe, will tend to form spherical micelles, with the hydrocarbon chains forming the core of the aggregate and the polar head groups arrayed at the particle surface.

KeyPoint

- The relative cross-sectional areas of a lipid's polar head group and hydrocarbon chain(s) determine the type of aggregate that it will form.

In organic media, lipids that have the shape of an inverted cone, i.e. that have a polar head group with a cross-sectional area that is smaller than the cross-sectional area of their hydrophobe, will tend to form reverse micelles, with the polar head groups forming the core of the aggregate and the lipid hydrophobes arrayed at the particle surface.

In aqueous media, lipids that have a cylindrical shape, i.e. that have a polar head group and hydrophobe of comparable cross-sectional area, will tend to form monolayers, and these can then pair together to form bilayers.

Cell membranes have a core structure of a lipid bilayer, associated with which there are proteins, glycoproteins and sterols.

Steroids

Steroids are based on a common (17-atom, *gonane*) molecular framework involving four fused cycloalkane rings – three cyclohexane rings and one cyclopentane ring.

The International Union of Pure and Applied Chemistry (IUPAC)-recommended ring-labelling and atom-numbering scheme is shown in Figure 12.5.

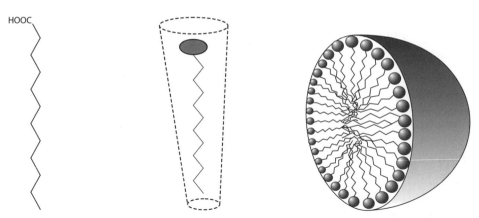

Fatty acid (left) has a truncated conical shape (centre) and so aggregates in aqueous media to form spherical micelles (shown in cut-away view, right).

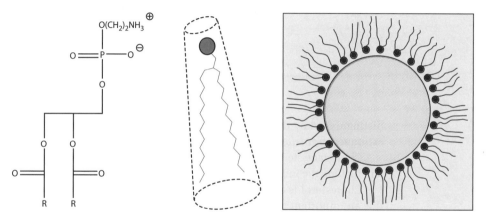

Phosphatidylethanolamine (left) has an inverse conical shape (centre) and so aggregates as reverse micelles in organic media (shown in cross-section, right)

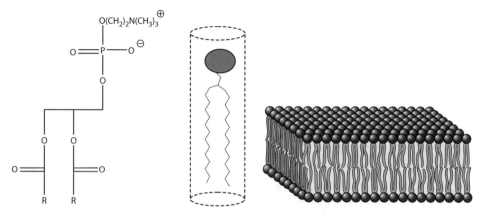

Phosphatidylcholine (left) has a roughly cylindrical shape (centre) and so aggregates to form monolayers and bilayers (the latter shown schematically, right).

Figure 12.4 Shapes of lipids and their aggregates.

Steroid nucleus ring labelling and atom numbering scheme.

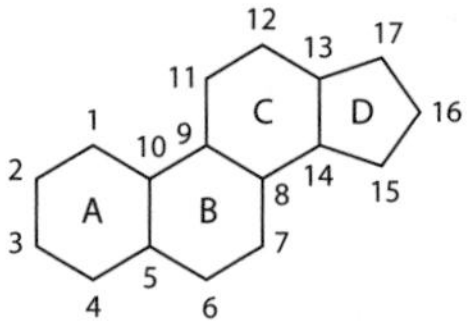

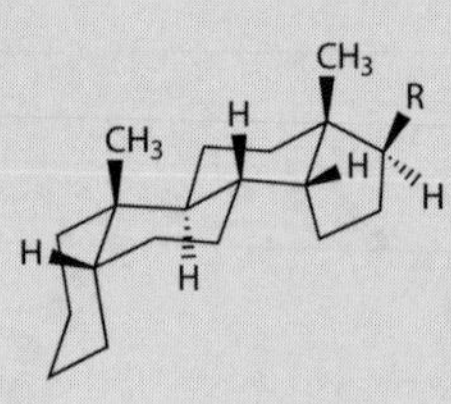

Atoms and functional groups that are attached to the steroid nucleus may be oriented above the plane of the ring to which they are attached, in which case they are referred to as β-substituents, or they may lie below the plane of the ring, in which case they are referred to as α-substituents (Figure 12.6). In two-dimensional (2D) representations, the β-substituents are drawn with wedge bonds and the α-substituents are drawn with dashed bonds (Figure 12.6).

Tip

Despite the way in which they appear in the standard 2D representations of their chemical structure, steroid molecules are *not* flat:

Classes of steroid

There are five different classes of steroid, distinguished by the nature of their molecular skeleton: *cholestanes, estranes, androstanes, pregnanes* and *cholanes*.

Cholestane steroids are based on the 27-carbon molecule cholestane and include the infamous steroid, *cholesterol* (Figure 12.7).

Cholesterol is an important component of cell membranes; it is an important determinant of cell membrane fluidity and permeability.

Figure 12.6
Steroid substituents. The methyl groups attached at C10 and C13, and the side chain, R, attached at C17, are all β-substituents. The hydrogen atoms attached at C9 and C15 are α-substituents.

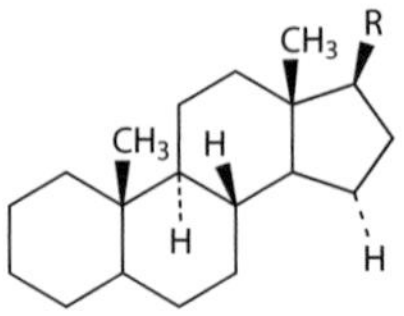

Figure 12.7
Cholestane (left) and cholesterol (right).

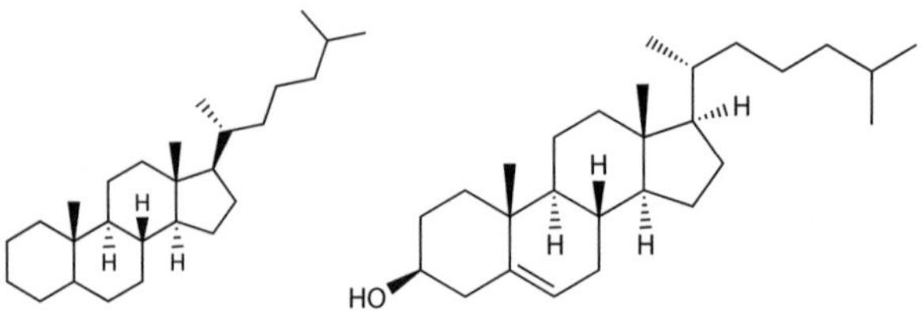

Cholesterol is also important as a precursor of bile acids, steroid hormones and vitamin D.

Cholesterol is absent in bacterial membranes, and it is replaced in plants by other sterols such as *stigmasterol*, and in fungi by sterols such as *ergosterol*.

Cholesterol is imported into cells from the blood, in the form of *low-density lipoprotein (LDL) complexes* (involving cholesterol, phospholipids, triglycerides and proteins).

High levels of blood cholesterol increase the risk of coronary heart disease and disease of the arteries.

Most foods actually contain very little cholesterol. Rich sources of cholesterol include eggs, offal and shellfish. Saturated fats in foodstuffs can be converted by the liver into cholesterol.

LDL – 'bad cholesterol' – takes cholesterol from the liver to the cells. If supply exceeds demand, this can cause a harmful build-up of cholesterol.

High-density lipoprotein (HDL) – 'good cholesterol' – takes cholesterol from the cells back to the liver, where it is broken down or excreted.

National Institute for Health and Care Excellence (NICE) and Department of Health guidelines are as follows:

- Total cholesterol < 5.0 mmol.L^{-1}
- LDL cholesterol < 3.0 mmol.L^{-1}

Estrane steroids are based on the 18-carbon molecule oestrane and include the female sex hormone, *oestradiol* (Figure 12.8).

The IUPAC systematic name of oestradiol is estra-1,3,5(10)-triene-3,17β-diol.

Oestradiol is the principal female sex hormone. It is responsible for the development and maintenance of the female sex characteristics, and has major effects on brain functioning and bone metabolism.

Androstane steroids are based on the 19-carbon molecule androstane and include the male sex hormone, *testosterone* (Figure 12.9).

Tip

Ethinyloestradiol (shown below) is a derivative of oestradiol that is used in the oral contraceptive pill. It is well absorbed orally, whereas oestradiol is not.

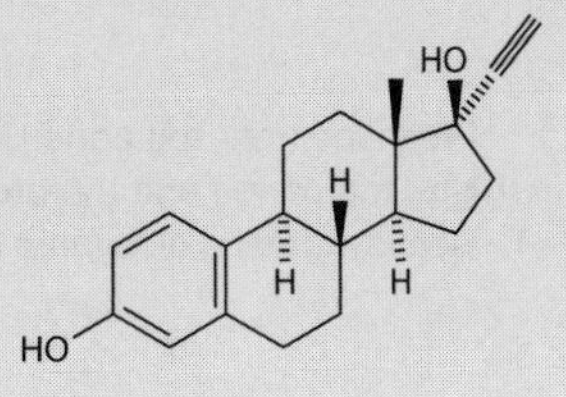

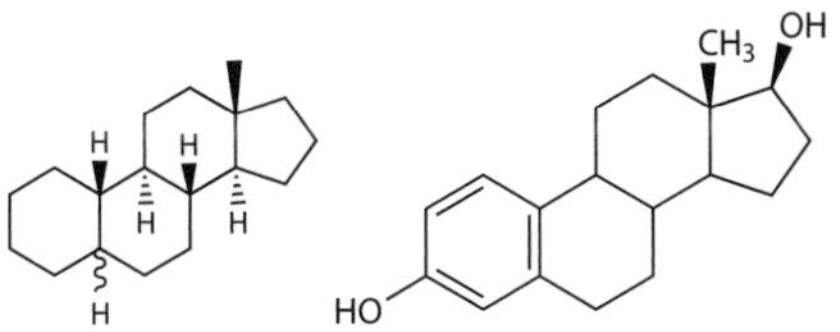

Figure 12.8
Oestrane (left) and 17-β-oestradiol (right).

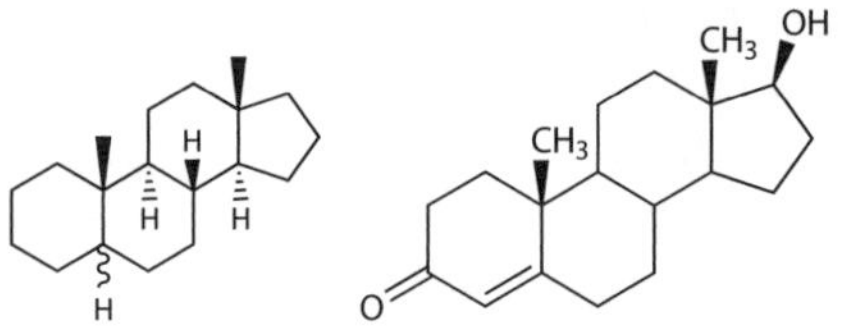

Figure 12.9
Androstane (left) and testosterone (right).

The IUPAC systematic name of testosterone is 17α-hydroxyandrost-4-en-3-one.

Testosterone is the male sex hormone. It is synthesised in the testes and is responsible for the development and maintenance of the male sex characteristics. It is synthesised from progesterone.

Tip

Epitestosterone is the naturally occurring inactive isomer of the hormone testosterone. The ratio of testosterone:epitestosterone in the human body (the T/E ratio) averages around 1.15:1. Athletes who use testosterone to try and enhance their athletic performance will typically have T/E ratios of 4:1 or even higher.

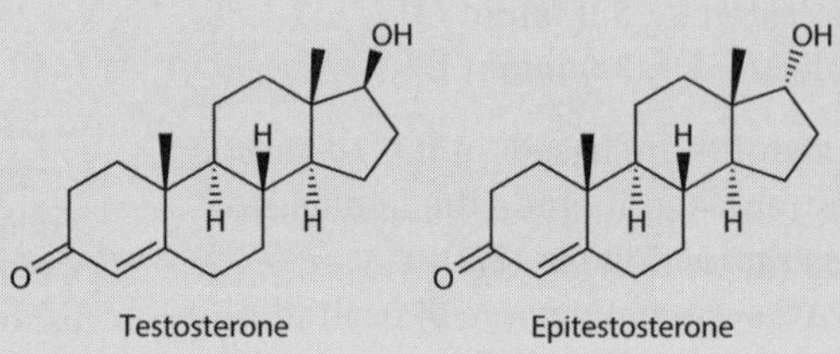

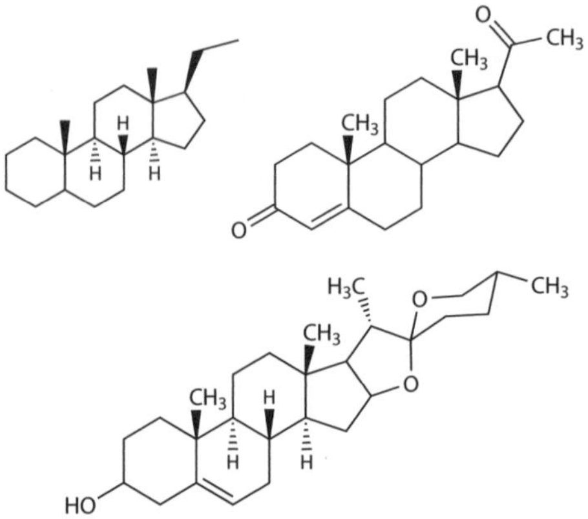

Testosterone   Epitestosterone

Figure 12.10
Pregnane (left, top panel), progesterone (right, top panel) and the precursor for its semisynthesis, diosgenin (lower panel).

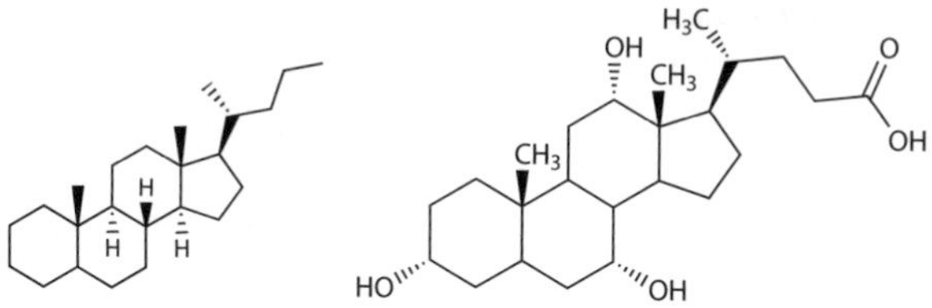

Figure 12.11
Cholane (left) and cholic acid (right).

Testosterone is an *anabolic* steroid (that is, it acts to stimulate intracellular protein production). It can be used by athletes to improve performance but is considered as a form of doping in most sports.

Pregnane steroids are based on pregnane and include the female sex hormone, *progesterone* (Figure 12.10).

The IUPAC systematic name of progesterone is pregn-4-ene-3, 20-dione.

Progesterone is produced in the corpus luteum, and is responsible for the changes that take place in (the luteal phase of) a woman's menstrual cycle. It also controls differentiation in the mammary glands.

Clinically, progesterone is administered to support pregnancy following *in vitro* fertilization (IVF).

Progesterone is manufactured commercially via semisynthesis from the steroid, diosgenin (obtained from yams: Figure 12.10).

Cholane steroids are based on the 24-carbon molecule cholane and include the bile acid, *cholic acid* (Figure 12.11).

Cortisol

The hormone cortisol (also known as hydrocortisone: Figure 12.12) is the principal glucocorticoid in humans. It is synthesised in the adrenal glands and is involved in the regulation of carbohydrate metabolism, and in stress management. It also has many effects on the immune system, and is used as an anti-inflammatory agent.

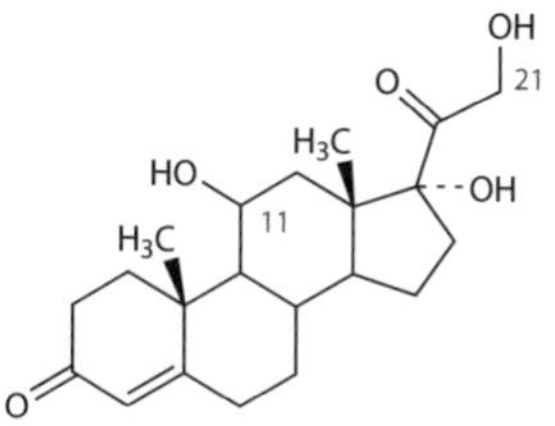

Figure 12.12
Cortisol.

11β, 17,21 -trihydroxypregn-4-ene-3,20-dione

Figure 12.13
Bile acids.

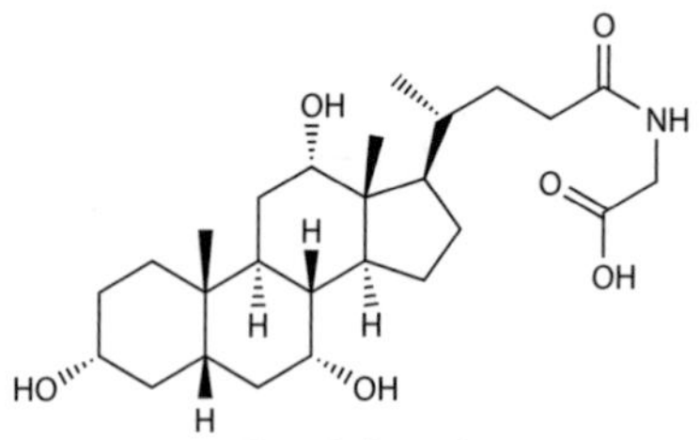

Glycocholic acid

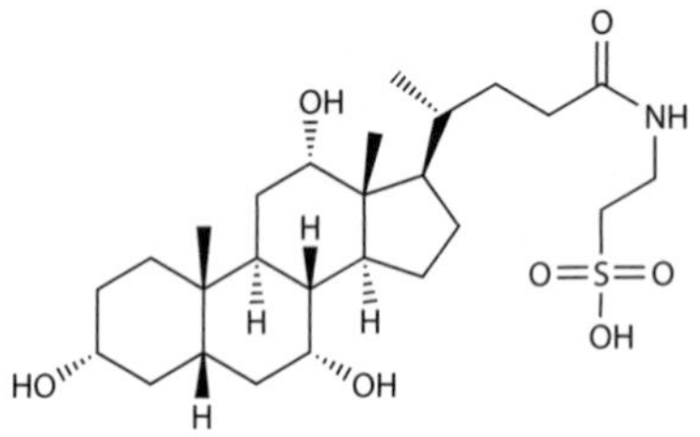

Taurocholic acid

Bile acids

Bile acids, found in bile in the form of bile salts, are used to facilitate the formation of lipid micelles within the small intestines, to promote lipid digestion (Figure 12.13). The principal species in human bile are the sodium salts of glycocholic acid and taurocholic acid. The bile acids are synthesised from cholesterol in the liver.

Cardiac glycosides

Cardiac glycosides are drugs used in the treatment of cardiac arrhythmia and congestive heart failure.

They have a steroid nucleus with an oligosaccharide head group attached at the 3-hydroxyl on the A-ring.

The cardiac glycoside digoxin (Figure 12.14) is obtained from the foxglove plant, *Digitalis lanata*.

Figure 12.14
Digoxin.

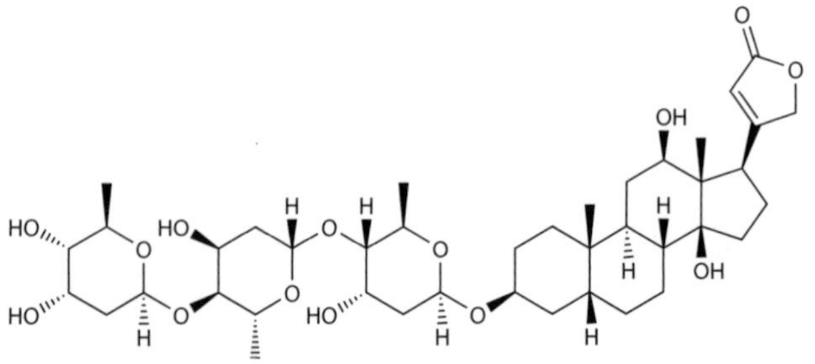

Self-assessment

1.

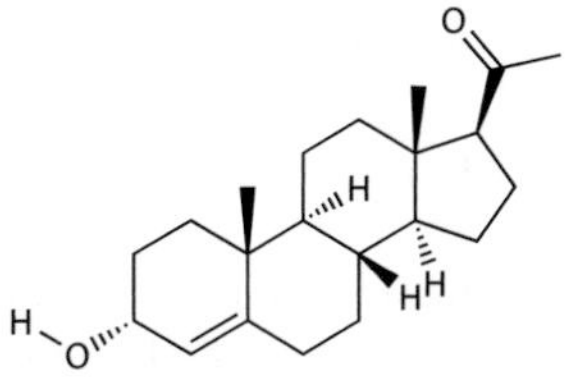

The steroid shown above is based on the skeleton:
a.   cholestane
b.   pregnane
c.   androstane
d.   oestrane

2. **Cholesterol:**
a.   is present in high levels in prokaryotes like bacteria
b.   is based on the cholestane skeleton
c.   has an 18-carbon skeleton
d.   is an anabolic steroid

3. **Ceramides:**
a.   are a type of glycolipid
b.   have polar head groups that involve one or more N-acetylneuraminic acid residues
c.   involve only a single hydrocarbon chain in their hydrophobe
d.   are based on sphingosine

4. $H_3C\text{-}CH_2\text{-}(CH=CH\text{-}CH_2)_3\text{-}(CH_2)_6\text{-}COOH$

The fatty acid shown above:

a.   has a melting point of 54°C
b.   has the shorthand code $\Delta^9 C_{18:3}$
c.   is an omega-3 fatty acid
d.   is palmitic acid

5.   **The melting point of a fatty acid:**
a.   increases with increasing unsaturation in its hydrocarbon chain
b..   decreases with increasing saturation in its hydrocarbon chain
c.   increases with increasing chain length in its hydrocarbon chain
d.   strongly depends on the nature of its polar head group

6.   **Glycerol:**
a.   is a type of steroid
b.   can react with fatty acids to form triglycerides
c.   provides the backbone structure of sphingolipids
d.   is a monohydric alcohol

7. **Progesterone:**
a. is an anabolic steroid
b. in males, is produced in the testes
c. is manufactured by semisynthesis from testosterone
d. is used to support pregnancy following *in vitro* fertilisation

8. **A lipid that has a polar head group with a cross-sectional area of 50 Å^2 and a hydrophobe with a cross-sectional area of 21 Å^2 is likely to aggregate in aqueous media to form:**
a. spherical micelles
b. monolayers
c. bilayers
d. reverse micelles

9. **Bacterial membranes:**
a. contain lipids with a high proportion of polyunsaturated fatty acids
b. have a high cholesterol content
c. contain lipids with branched-chain fatty acids
d. are rich in lipids containing linolenic acid

10. **Lecithins are otherwise known as:**
a. sphingomyelins
b. cardiolipins
c. glycerosphingolipids
d. glycerophospholipids

chapter 13
Chemical stability of drugs

Overview

After learning the material presented in this chapter you should:

- understand the factors that affect a drug's shelf-life stability and its stability after administration
- understand the chemistry involved in drug stability
- know and be able to discuss specific examples of drug instability
- know and be able to discuss strategies to improve drug shelf-life stability
- know and be able to discuss strategies to improve drug stability after administration
- understand and be able to explain prodrug strategies to improve therapy.

Types of drug stability

- There are two types of drug stability to consider:
  1. shelf-life stability
  2. stability of a drug after administration.

- Many of the factors that affect the shelf-life stability of a drug will also affect the stability of the drug after administration.
- The key factors affecting drug stability include:
  1. Moisture (hydrolysis): The hydrolysis of drugs can be a problem both before and after administration. Moisture also promotes microbial growth, further affecting shelf-life.
  2. pH: Many drugs are sensitive to acidic or basic conditions and small changes in pH can lead to significant decomposition. At pH 2–3 aspirin is stable, although outside this window decomposition occurs; at pH 10 aspirin is rapidly hydrolysed.
  3. Oxygen (oxidation): In the presence of oxygen certain functional groups within a drug molecule can be oxidised; both captopril and chlorpromazine undergo oxidation. After administration the liver will begin to oxidise drug molecules to facilitate excretion.

KeyPoints

- Shelf-life is defined as the time taken for a drug's pharmacological activity to decline to an unacceptable level.
- Different dosage forms of the same drug will have different shelf-lives.

4. Temperature: Higher temperatures accelerate chemical processes and will increase the rate of drug hydrolysis, oxidation and reduction.
5. Light (photolysis): Exposure to daylight can significantly reduce the shelf-life of a drug. Sodium nitroprusside must be protected from daylight during administration as it rapidly decomposes within a few hours of exposure.
6. Racemisation: Where a drug is used as a single enantiomer it may be possible to racemise before or after administration. Racemisation may lead to the formation of an inactive enantiomer or an enantiomer with a negative side effect profile, for example thalidomide.
7. Dosage form: Solid dosage forms will be more stable than liquid dosage forms.
8. Drug incompatibility: Reactions between different components of a formulation may reduce the shelf-life of a particular formulation.

Functional group reactivity

■ When considering drug stability it is important to consider the individual functional groups present in the drug molecule and the environment they find themselves in.
■ A drug may be stable once in the blood stream but be unstable in the highly acidic environment of the stomach.
■ The intrinsic reactivity of a functional group does not change when incorporated into a drug.
■ Reactivity may be modified in the drug but an ester will always be an ester and will react as such (Figure 13.1).
■ Understanding the reactivity of individual functional groups provides a valuable insight into how a drug molecule containing multiple functional groups will behave (Figures 13.2–13.6).
■ The features of a drug that provide the therapeutic effect will sometimes also be responsible for the instability of the drug.

Figure 13.1
Reactivity of functional groups.

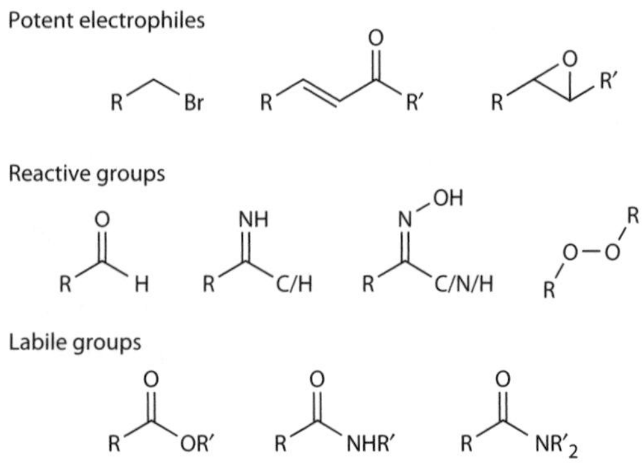

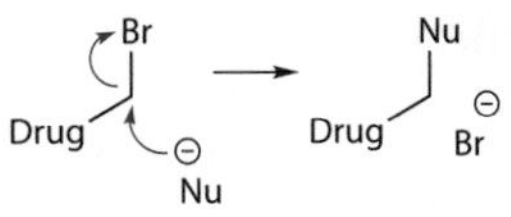

Figure 13.2
Reactivity of alkyl halides.

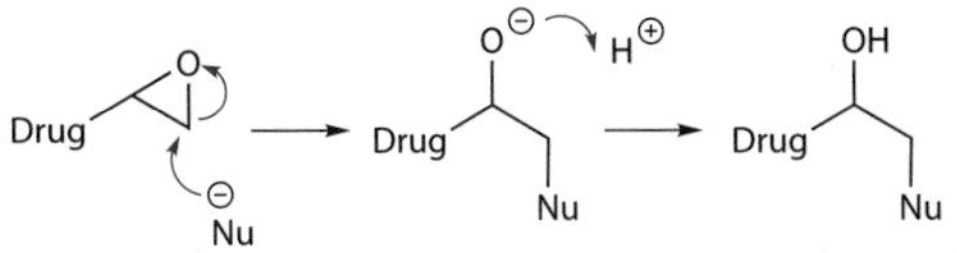

Figure 13.3
Reactivity of epoxides.

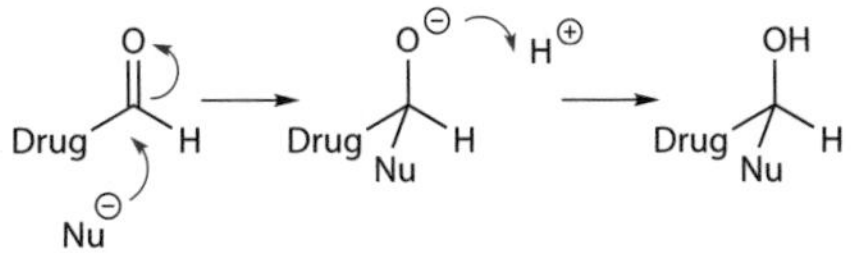

Figure 13.4
Reactivity of carbonyl groups (1,2-addition).

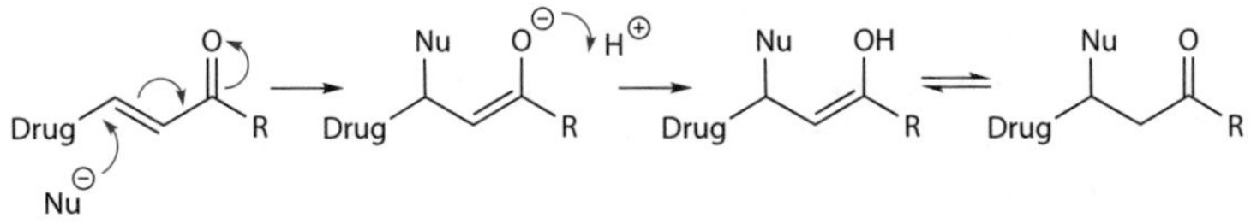

Figure 13.5
Reactivity of carbonyl groups (1,4-addition).

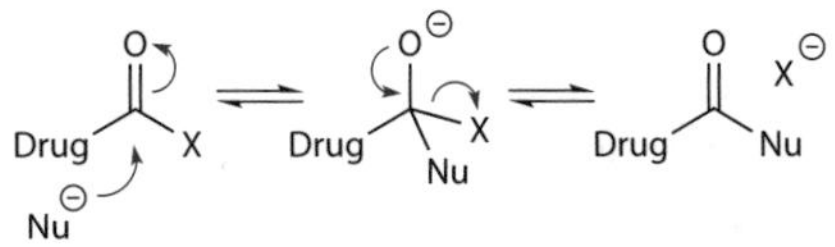

Figure 13.6
Reactivity of carboxyl groups.

- Aspirin inhibits the cyclooxygenase (COX) enzyme through the irreversible acetylation of a serine residue. The covalent attachment of an acetyl group in the active site prevents prostaglandin production (Figure 13.7).
- The chemistry involved for the serine acetylation is similar to a transesterification reaction.
- The reactivity of aspirin towards this addition–elimination chemistry will also make it reactive towards related reactions, including ester hydrolysis.
- If aspirin were more stable towards ester hydrolysis it would not be able to inhibit prostaglandin production. There are times when chemical instability is unavoidable and instead must be managed.

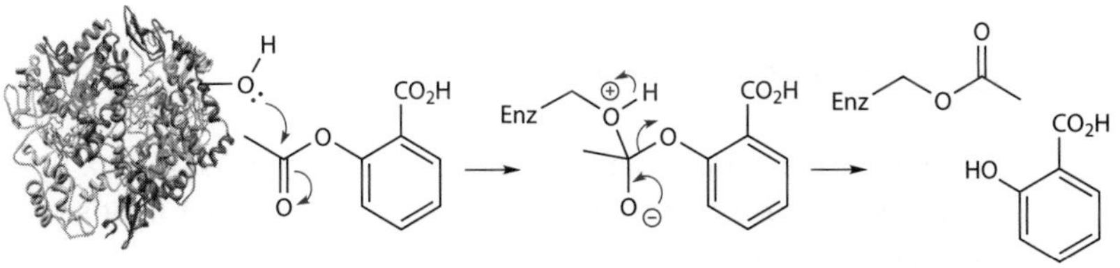

Figure 13.7 Mode of action of aspirin.

Strategies to improve shelf-life stability

- Practical steps can be taken to increase the shelf-life of a drug.
- Moisture and hence hydrolysis can be minimised through the use of airtight lids.
- Drugs should be formulated at an appropriate pH using buffers where necessary to maintain a pH suitable for that particular drug.
- Oxidation can be minimised by filling bottles to have the minimum possible air pocket at the top, packaging under nitrogen or carbon dioxide and using antioxidant excipients. Exposure to heavy metals that would promote free radical oxidation should also be avoided.
- Storage and transportation at an appropriate temperature will limit thermal decomposition and extend the shelf-life of unstable drugs. Refrigeration may be necessary.
- Drugs that are light-sensitive should be stored in amber-coloured bottles to prevent photolysis and free radical degradation.
- For drugs that are racemised through either an acid- or base-catalysed mechanism pH control will be essential.
- Solid dosage forms are more stable than liquid dosage forms. In a solid, individual molecules are rigidly held in place; in solution, molecules are colliding with each other and are able to react with traces of water more readily.
- Microbial degradation can be avoided by preparing and storing the dosage form under sterile conditions. The addition of antimicrobial excipients will also extend shelf-life.

KeyPoints

- The body will start metabolising a drug immediately after administration. The drug cannot be metabolised too quickly.
- Once a drug has carried out its function, it must be excreted.
- A drug cannot be so stable that it is allowed to accumulate within the body.

Stability after administration

- A drug will only be effective if, after administration, it is stable enough to reach its target site in sufficient concentration to bring about the desired therapeutic effect.
- The features which make a molecule reactive in a flask will also make it reactive in the body.
- The same chemistry is involved, just different 'reagents'.

Amide and ester hydrolysis

- Both esters and amides are susceptible to hydrolysis *in vivo*.
- Protease and esterase enzymes cleave amide and ester bonds in drug molecules.

- The greater electron delocalisation within the amide bond will make amides more stable towards enzymatic hydrolysis than the analogous ester.
- Procainamide undergoes hydrolysis at a slower rate than procaine (Figure 13.8).
- A drug that contains both ester and amide functional groups will initially undergo hydrolysis at the more reactive ester.
- Propanidid will undergo selective hydrolysis at the ester group in the presence of the amide. The reaction is chemoselective (Figure 13.9).

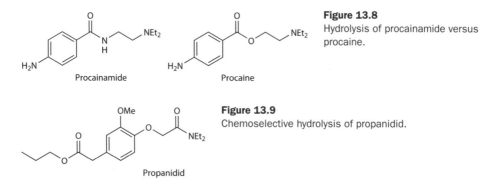

Figure 13.8
Hydrolysis of procainamide versus procaine.

Procainamide

Procaine

Figure 13.9
Chemoselective hydrolysis of propanidid.

Propanidid

Hydrolysis of enantiomers

- The individual amino acids from which proteins are composed are chiral; the proteins by extension are therefore chiral environments.
- Proteins will not necessarily interact with the individual enantiomers of a drug in the same way.
- One enantiomer of a drug may be recognised by a protein whereas its corresponding enantiomer is not.
- For the two enantiomers of a drug, it may be that only one enantiomer will be metabolised *in vivo*.
- The two enantiomers of prilocaine have comparable local anaesthetic activity; however, the amide bond of the *R*-enantiomer is selectively hydrolysed to toluidine which can lead to methaemoglobinaemia, a condition where there is a reduced ability to release oxygen to the tissues.
- The *S*-enantiomer of prilocaine is not hydrolysed *in vivo* (Figure 13.10).

KeyPoints

- Although two enantiomers can be distinguished *in vivo* and undergo selective reactions, in the laboratory the two enantiomers would react identically.
- Both enantiomers of prilocaine would undergo hydrolysis in the laboratory.
- Enantiomers can only be distinguished in a chiral environment.

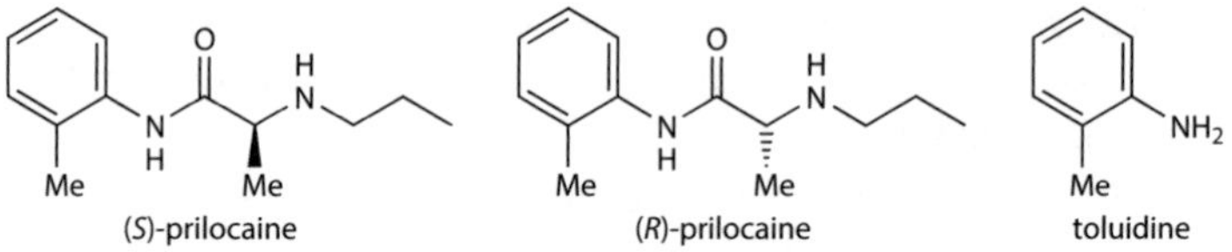

Figure 13.10
Selective hydrolysis
in prilocaine.

(S)-prilocaine   (R)-prilocaine   toluidine

Strategies to improve drug stability after administration

Modify the structure of the drug

■ When a drug contains a reactive functional group it may be possible to modify that group and retain biological activity.
■ Pilocarpine is used in the treatment of glaucoma. However, the cyclic ester is particularly prone to hydrolysis and this results in a short *in vivo* half-life (Figure 13.11).
■ The structure of pilocarpine has been modified to reduce the electrophilicity of the ester towards hydrolysis while retaining biological activity (Figure 13.12).
■ Replacement of the ester with a carbamate increases charge delocalisation since the nitrogen atom is able to donate electron density into the carbonyl group in addition to the electron donation from the oxygen atom.

■ The carbamate shares many of the structural features of the ester and the modified pilocarpine analogue is able to interact with the same receptor as pilocarpine as a result.
■ The ester and carbamate groups are said to be bioisosteres.

Tip

Bioisosteres are substituents or groups with similar chemical or physical properties which produce broadly similar biological properties when incorporated into a molecule.

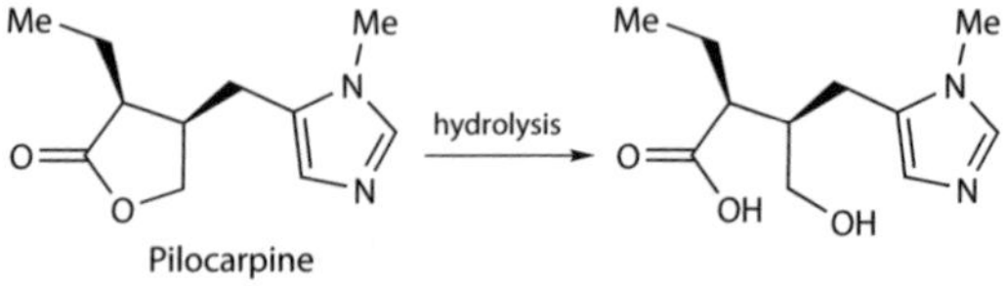

Figure 13.11
Pilocarpine hydrolysis.

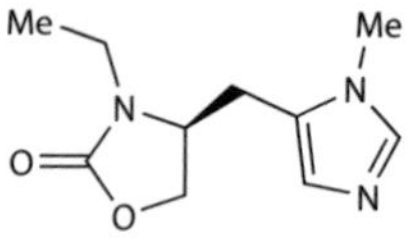

Pilocarpine

Figure 13.12
Modified pilocarpine structure.

Administration as a prodrug

■ If a drug is being rapidly metabolised or not being absorbed or distributed efficiently following administration it may be possible to modify the drug using a prodrug strategy.

- A prodrug is a drug precursor that will improve the bioavailability of a pharmacologically active drug while not being pharmacologically active itself.
- It is only upon metabolism that the prodrug will be converted to the pharmacologically active drug.
- A prodrug strategy will enable a lower dose of a poorly bioavailable drug to be administered as there will be a more efficient delivery of the active drug to the site of action.
- Enalaprilat is an angiotensin-converting enzyme (ACE) inhibitor with poor oral bioavailability due to the presence of two carboxylic acid groups. Adopting a prodrug strategy, enalapril was developed (Figure 13.13).

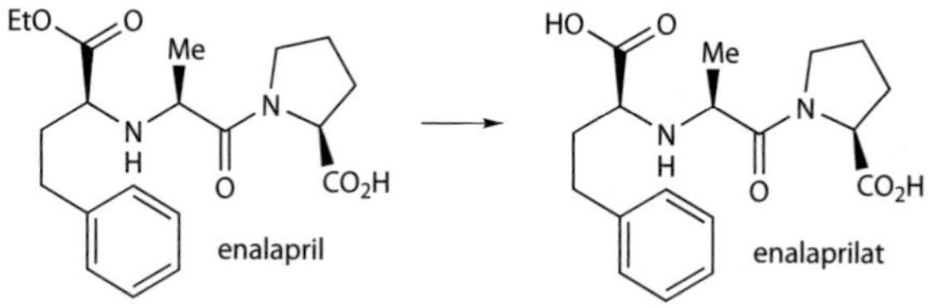

Figure 13.13
Hydrolysis of the enalapril prodrug.

- One of the carboxylic acid groups is masked as an ester to aid absorption.
- The labile nature of the ester group is utilised to release the active drug once the prodrug has been delivered into the blood stream.
- The antibiotic tizoxanide is administered as the prodrug nitazoxanide. Hydrolysis of the ester releases the pharmacologically active drug (Figure 13.14).
- Both carboxylic acids and hydroxyl groups on drug molecules can be masked as a prodrug.

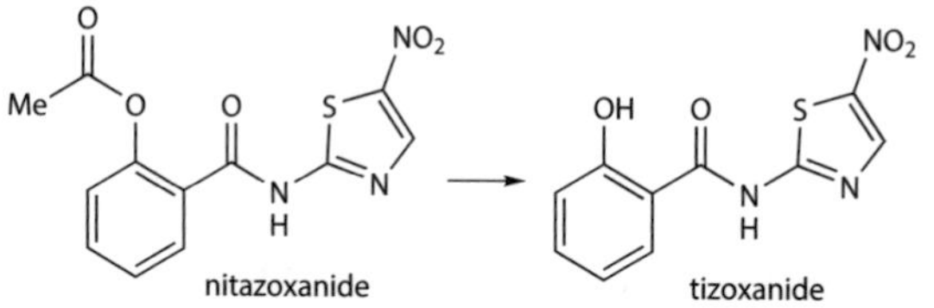

Figure 13.14
Hydrolysis of the nitazoxanide prodrug.

Modify the route of administration or dosage form

- The highly acidic environment of the gastrointestinal tract can lead to the hydrolysis of a drug and result in poor oral bioavailability.
- An alternative route of administration may be required that avoids the acidic environment of the stomach. Options include:
  1. intravenous (into a vein) administration
  2. intramuscular (into a muscle) administration
  3. intranasal (through the nose) administration
  4. subcutaneous (under the skin) administration
  5. transdermal (through the skin) administration.

■ Alternatively the drug formulation might be modified to make use of an enteric coating to allow oral administration.

■ The enteric coating is stable at low pH and protects the drug in the stomach.

■ The coating is less stable at higher pH and will begin to dissolve during transit through the more basic environment of the small intestine.

■ The drug is delivered unchanged to the small intestine for absorption.

■ Complexation of an active drug can reduce the rate of both hydrolysis and oxidation.

■ Caffeine–benzocaine complexes, for example, are less prone to ester hydrolysis (Figure 13.15).

Figure 13.15
Caffeine–benzocaine complexes
demonstrate enhanced stability.

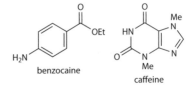

benzocaine

caffeine

Self-assessment

1. **Which of the following strategies would not extend the shelf-life of a drug?**
a. formulating with an antimicrobial excipient
b. replacing the air in the package with nitrogen
c. formulating with an enteric coating
d. formulating with an antioxidant

2. **Enalapril is converted to enalaprilat *in vivo* by which of the following reactions?**
a. esterification
b. hydrolysis
c. oxidation
d. reduction

3. **Which of the functional groups labelled below would be responsible for making the drug candidate susceptible to oxidation?**

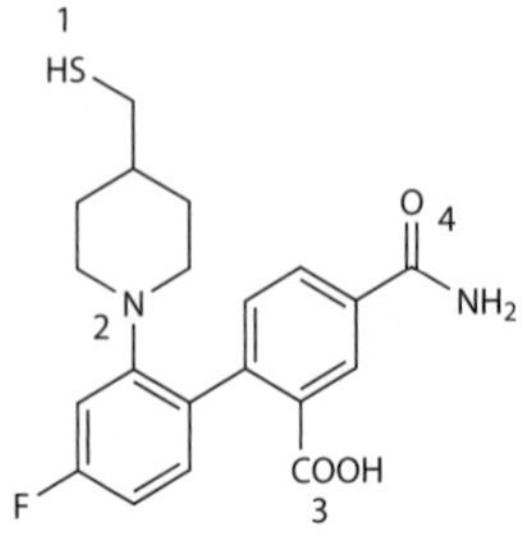

a. functional group 1
b. functional group 2
c. functional group 3
d. functional group 4

4. Which of the following observations would indicate that aspirin has gone off?
a. Tablets taste sour.
b. Tablets smell of rotten eggs.
c. Tablets taste sweet.
d. Tablets smell of vinegar.

5. Which of the following drugs could be susceptible to racemisation?

a.

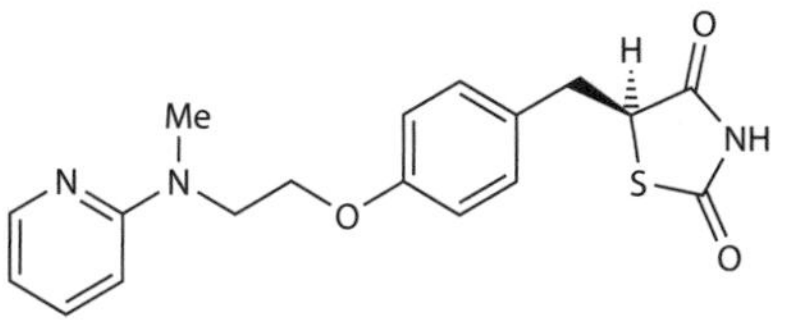

b.

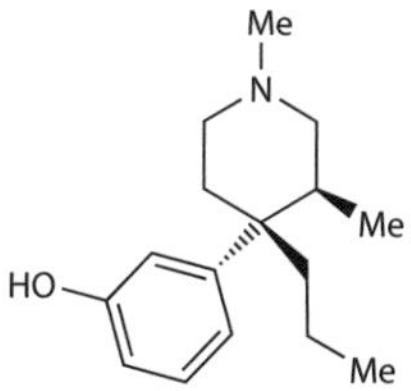

c.

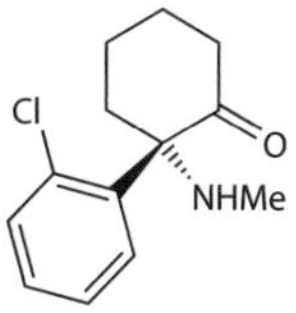

d.

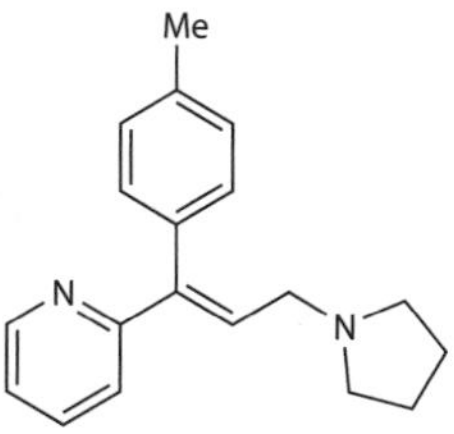

6. Which of the following strategies would not extend the shelf-life of a drug that is susceptible to oxidation?
a. refrigerating the formulation
b. protecting the formulation from light
c. replacing the air in the package with nitrogen
d. adding a heavy metal to the formulation

7. **Which of the following strategies would be most effective in extending the shelf-life of a light-sensitive drug?**
a.   formulation with an antioxidant
b.   replacing the air in the package with carbon dioxide
c.   packaging in amber bottles
d.   storage in a concentrated stock solution

8. **Which of the following should specifically be avoided while taking an enteric-coated drug formulation?**
a.   antacids
b.   red meat
c.   grapefruit juice
d.   alcohol

9. **Which of the following molecules is a prodrug?**
a.

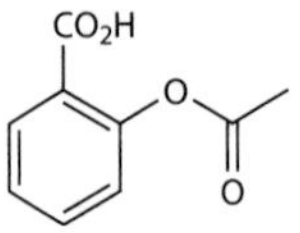

b.

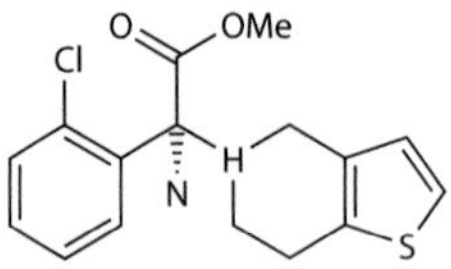

c.

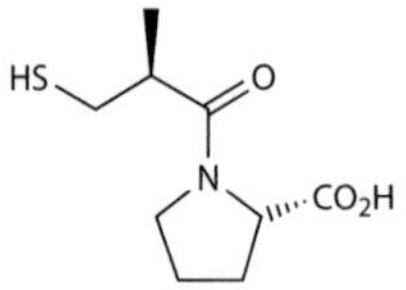

d.

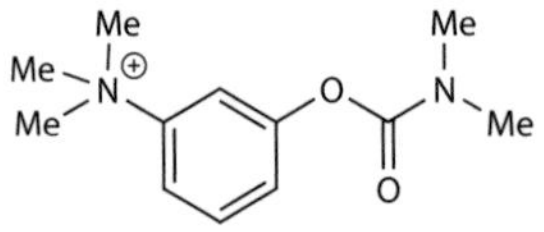

10. **Which of the functional groups labelled below would be responsible for making the drug candidate susceptible to hydrolysis?**

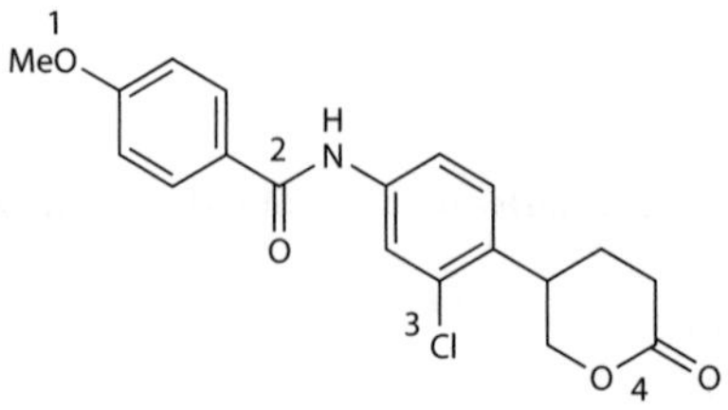

a.   functional group 1
b.   functional group 2
c.   functional group 3
d.   functional group 4

chapter 14
Drug metabolism

Overview

After learning the material presented in this chapter you should:

- understand and be able to explain the significance of drug metabolism
- be able to distinguish between phase I, phase II and phase III metabolic processes
- appreciate the significance of the liver in drug metabolism and understand the first-pass effect
- be able to study a drug structure and predict its possible metabolites
- appreciate the various physiological and pathological factors that can affect the metabolism of drugs
- understand and be able to explain why the genetic variation (polymorphism) of drug-metabolising enzymes can lead to variation in the efficacy of drug treatments between individuals.

- Drug metabolism is simply a form of *xenobiotic metabolism*. It involves the *biotransformation* of drug substances via the metabolic pathways that the body has evolved to detoxify and eliminate (foreign) compounds that are likely to be harmful.
- The lipid bilayers of cell membranes provide an effective barrier against the unwanted entry of polar/hydrophilic compounds into cells, but non-polar/lipophilic compounds can cross the cell membrane and readily enter cells. The biotransformation reactions involved in xenobiotic metabolism have thus evolved to deal (primarily) with foreign lipophilic compounds.
- Some pharmaceuticals act as substrates or inhibitors of the enzymes involved in xenobiotic metabolism and this is a frequent cause of hazardous *drug interactions*.
- In some cases, the metabolic products generated through the biotransformation of pharmaceuticals can actually cause toxic effects.
- The enzymes responsible for the biotransformation of drugs show genetic variation from one individual to another and this is why some drugs can be effective in treating some patients but not others.

Phases of metabolism

It is commonly considered that there are two *phases of drug metabolism*, referred to quite simply as phase I and phase II.

Phase I metabolism

- Phase I metabolism involves the *modification* of xenobiotic compounds through the introduction of reactive and/or more polar groups via oxidation, reduction and hydrolysis reactions. Most phase I drug metabolism is performed in the liver and involves the haem-containing *cytochrome P450*-dependent mixed function oxidase system.
- Cytochrome P450 enzyme systems use oxygen to introduce a hydroxyl group into the drug structure or else cause the dealkylation of N-, S- or O-alkyl groups (Figure 14.1). The reactions catalysed by the cytochrome P450 enzymes require oxygen and the coenzyme NADPH. The NADPH is oxidised to NADP$^+$ in the reaction.
- The genes that code for the cytochrome P450 enzymes, and the enzymes themselves, are designated by the abbreviation CYP.

Figure 14.1
Examples of phase I drug metabolism.

N-dealkylation of imipramine

Aromatic hydroxylation of phenytoin

Aliphatic hydroxylation of ibuprofen

N-oxidation of olanzapine

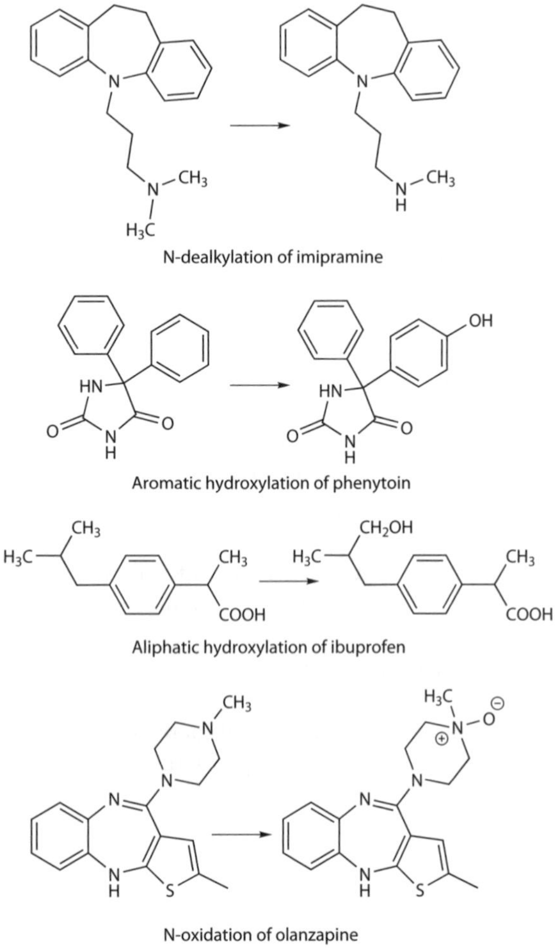

The genes are named using italics, and the associated protein is named using normal characters. For example, *CYP2C9* is the gene that codes for the enzyme CYP2C9.

■ In CYP nomenclature, the first digit specifies the protein family, the following letter specifies the subfamily and the final digits specify the particular protein. CYP3A4 is thus a CYP in family 3, subfamily A, member 4.

■ CYPs are located primarily in the membranes of the endoplasmic reticulum in the cell.

■ Mutations in the genes that code for CYPs have given rise to different variants of the enzymes. These variants have different amino acid sequences. The individual variants are referred to as *isoenzymes*.

■ Other phase I enzyme systems that catalyse biotransformations involving oxidation/reduction reactions include: monoamine oxidase (MAO), flavin-containing monoamine oxidase (FMO), peroxidase and alcohol dehydrogenase.

■ Phase I enzymes that lead to the hydrolysis of drug substances include esterases and amidases.

■ If the metabolites of phase I reactions are sufficiently polar, they may be readily excreted without further processing. Many phase I drug metabolites, however, are not eliminated rapidly and undergo a subsequent reaction in which an endogenous substrate is combined with the newly incorporated functional group to form a highly polar conjugate.

Phase II metabolism

■ Phase II metabolism involves the *conjugation* of the phase I metabolites to other polar compounds. The phase I activated xenobiotic metabolites are conjugated with charged species such as glutathione (GSH), sulfate, glycine or glucuronic acid (Figure 14.2).

■ The sites on drugs that are susceptible to phase II conjugation reactions include carboxyl, hydroxyl, amine and sulfhydryl groups.

■ Phase II metabolic reactions are catalysed by broad-specificity *transferases* which, working in combination, can metabolise almost any lipophilic compound that contains nucleophilic and/or electrophilic groups. One of the most important phase II enzymes is that of glutathione-*S*-transferase, which catalyses *glutathione conjugation*.

■ Other major classes of phase II enzymes include: *methyltransferases* (which catalyse *methylation*, and use the cofactor *S*-adenosyl methionine), *sulfotransferases* (which catalyse *sulfation*, and use 3'-adenosine 5'-phosphosulfate), UDP-glucuronosyltransferases (which catalyse *glucuronidation*, and use the coenzyme UDP-glucuronic acid).

Figure 14.2
Examples of phase II drug metabolism.

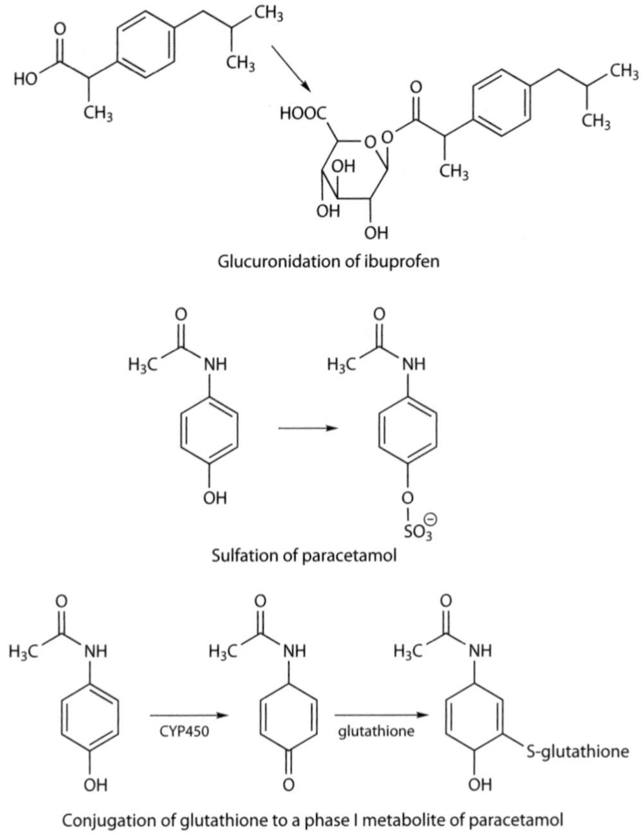
Glucuronidation of ibuprofen
Sulfation of paracetamol
CYP450
glutathione
S-glutathione
Conjugation of glutathione to a phase I metabolite of paracetamol

KeyPoints

- There are three phases of drug metabolism.
- Phase I metabolism involves oxidation, reduction or hydrolysis reactions, and results in the introduction of reactive and/or polar groups into a drug molecule.
- Phase II metabolism involves conjugation of a phase I metabolite to an endogenous polar molecule.
- Phase III metabolism involves processing of a conjugated drug metabolite and its efflux from a cell for excretion.

- Some authorities consider a further phase of drug metabolism: phase III metabolism.

Phase III metabolism

- Phase III metabolism involves further processing of the conjugated xenobiotic metabolites and the efflux of these from cells for excretion. The anionic groups on the phase III metabolites are recognised by specific *efflux proteins* – members of the *multidrug resistance protein* family – and these proteins catalyse the ATP-dependent transport of the compounds across the cell membranes.

Site of metabolism

Although every tissue in the body can metabolise drugs to some extent, it is generally the liver that provides the major site of their metabolism.

The liver's contribution to drug metabolism is significant partly because it is a large and highly vascular organ, but also because it has a high concentration of drug-metabolising enzymes relative to other organs, and because it is the first organ perfused by chemicals absorbed from the gastrointestinal tract following oral administration.

Drugs that are taken into the gastrointestinal tract and enter into the hepatic circulation through the portal vein are well-metabolised and are said to show the *first-pass effect.*

Factors affecting drug metabolism

- The extent of metabolism of a drug is influenced by physiological factors which include a patient's age, nutritional state, gender and pharmacogenetics profile, and also by pathological factors, and so are commonly affected, therefore, by liver, kidney and heart disease.
- In general, drugs are metabolised more slowly in fetal, neonatal and elderly humans.
- Geriatric patients very often require lower doses of drugs than younger adults firstly, because they have poorer blood circulation and/or reduced kidney function (which means that drugs are eliminated from the body less quickly); secondly, because of their increased levels of body fat (which means that drugs are accumulated more in the body); and thirdly, because they suffer reduced oxygen availability (which means that they have a reduced rate of phase I metabolic processes).

> **KeyPoints**
>
> - The extent of drug metabolism is affected by a patient's age, nutritional state, gender and genetic makeup …
> - … and also by any diseases they may have (in particular those affecting the liver and kidney).

- In the fetus and neonate, the liver is not fully developed and so drug metabolism is slower than in adults. CYP activity in neonates, for example, is generally 30–50% of that in a healthy adult. In consequence, the half-lives of drugs in neonates are ~10-fold longer than in adults.
- Adult patterns of CYP expression are attained in neonates by 3–6 months, and the full adult complement of phase I and phase II metabolic processes is developed in children by the age of 2–3 years.
- Genetic variation (polymorphism) accounts for some of the variability in the effects of drugs. With *N*-acetyltransferases (involved in phase II reactions), individual variation creates a group of people who acetylate slowly (*slow acetylators*) and another group who acetylate quickly. This variation may have dramatic consequences because the slow acetylators are more prone to issues arising through dose-dependent drug toxicity.

- Cytochrome P450 monooxygenase system enzymes (CYPs) can also vary across individuals, with deficiencies occurring in 1–30% of people, depending on their ethnic background.
- For drug metabolism, the most important polymorphisms are those of the genes coding for the CYPs: CYP2C9, CYP2C19, CYP2D6 and CYP3A4. Polymorphisms in these CYPs can result in therapeutic failure or severe adverse reactions.
- Specific CYP polymorphisms are identified numerically, such as CYP2D6*5, which indicates polymorphism 5 in CYP2D6. This polymorphism affects around 5% of the population, and corresponds to a gene deletion. Patients with the CYP2D6*5 polymorphism do not produce CYP2D6.
- Patients with the CYP2D6*2 polymorphism have a duplication of the gene for CYP2D6, and such patients are said to have the *ultra-rapid metaboliser* phenotype. The CYP2D6*2 polymorphism is prevalent in East Asians and people from equatorial Africa.
- The levels of expression of CYP proteins can be influenced by drugs and can also be affected by ingested foodstuffs.
- Some drug substances can cause *CYP induction*. That is, they lead to an increase in CYP expression. Drugs that cause induction of CYPs include the antibiotics rifampicin and erythromycin, and the anticonvulsant drug, phenytoin.
- Other drug substances can cause *CYP inhibition*, and may thus interfere with the metabolic processing of drugs that are subsequently administered. Drugs that act as CYP inhibitors include the histamine H_2-receptor antagonist, cimetidine, and the calcium channel blocker, diltiazem.
- *Carbon monoxide* is a CYP inhibitor: it attaches to the haem iron (in the Fe^{2+} state), prevents the haem group from binding oxygen, and so interrupts the enzyme's catalytic cycle and prevents formation of any oxidised drug metabolite.
- Herbal medicinal products that cause changes in CYP expression levels may give rise to problematic drug interactions. *St John's wort*, for example, a popular herbal product taken as a treatment for depression, causes significant induction of CYP3A4, and long-term use of St John's wort is shown to reduce the clinical efficacy of any drug that is a substrate for CYP3A4.
- Similar problems are also seen as a result of drug interactions with foodstuffs. *Grapefruit* and grapefruit juice, for example, interact with a large number of drugs, in many cases leading to adverse effects. The furanocoumarins in grapefruit inhibit CYP3A4, and drugs that are substrates of this enzyme are not metabolised as normal.
- Drug metabolic processes can sometimes cause the *bioactivation* of an administered drug, leading to the formation of metabolites that are (highly) reactive and give rise to toxic effects.

Self-assessment

1. Xenobiotic metabolic processes have evolved to ensure that the body is protected from:
a. highly polar compounds
b. pharmaceutical drugs
c. highly lipophilic compounds
d. drugs of abuse

2. Metabolites A and B (shown below left and centre) are formed through biotransformation of propranolol (shown below right).

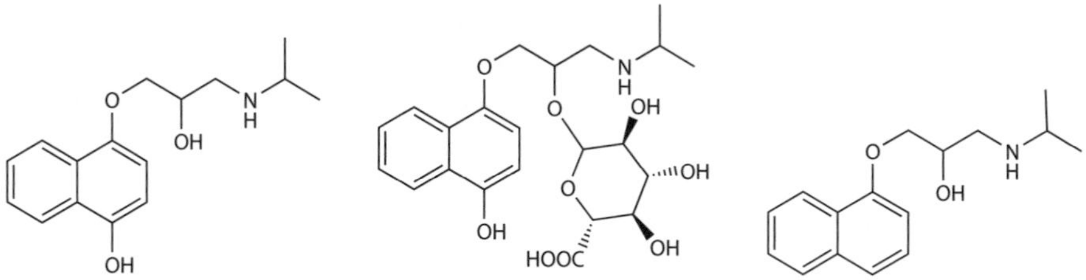

Metabolite A   Metabolite B   Propranolol

Identify the types of metabolism involved in production of A and B:

a. A: phase I / B: phase I
b. A: phase II / B: phase II
c. A: phase I / B: phase II
d. A: phase II / B: phase I

3. Drug metabolism in neonates:
a. is very much faster than in adults
b. does not involve cytochrome P450 enzymes
c. involves only phase II enzymes
d. leads to an increase in the half-life of drugs by comparison with their half-life in adults

4. Identify the enzyme in humans that is most likely to be responsible for the biotransformation shown below:

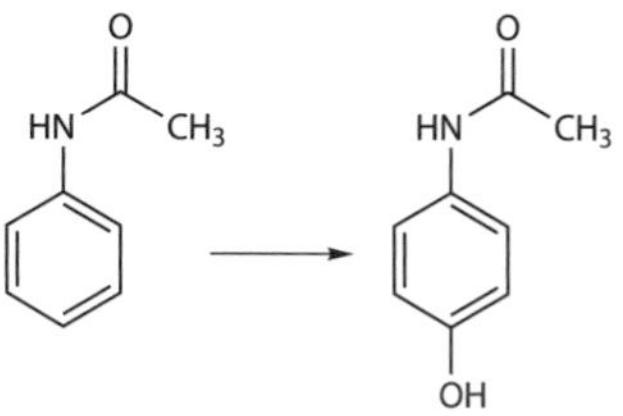

a. alcohol dehydrogenase
b. cytochrome P450
c. esterase
d. methyltransferase

5. **If a patient who is receiving a course of treatment of the anticoagulant drug warfarin is subsequently administered the drug rifampicin, the administered dose of warfarin has to be increased (in order to achieve the same anticoagulant effect). This is most likely necessary because:**
a. rifampicin is highly chemically reactive and destroys the warfarin
b. rifampicin causes increased absorption of the warfarin
c. rifampicin induces the enzyme that metabolises the warfarin
d. rifampicin inhibits the enzyme that metabolises the warfarin

6. **Some orally administered drugs can give rise to hepatotoxicity, and this is because:**
a. these drugs are rapidly absorbed in the mouth and then pass directly to the liver
b. these drugs are not metabolised by the liver
c. the liver is only poorly supplied with blood and the drugs cannot be metabolised quickly enough
d. the liver is exposed to high levels of these drugs and suffers if the drugs are subject to bioactivation

7. **Carbon monoxide :**
a. has no effect on CYP enzymes
b. acts to induce CYP enzymes
c. acts to inhibit CYP enzymes
d. is metabolised by CYP enzymes

8. **Phase II biotransformation reactions:**
a. generally lead to the inactivation of a drug
b. are catalysed by CYP enzymes
c. occur at the same rate in adults and neonates
d. lead to the formation of metabolites with decreased solubility in water

9. **The physiological changes that can adversely affect drug metabolism in geriatric patients include:**
a. increased renal function in these patients
b. reduced body fat in these patients
c. poor blood circulation in these patients
d. increased CYP activity arising from increased O_2 availability in these patients

10. **CYP (iso)enzymes are found in the:**
a. cell nucleus
b. cell membrane
c. endoplasmic reticulum
d. mitochondria

chapter 15
Molecular spectroscopy and pharmaceutical analysis

Overview

After learning the material presented in this chapter you should:

- know about the various regions of the electromagnetic spectrum, and the properties of infrared, ultraviolet (UV) and visible light
- be able to interconvert between electromagnetic wave frequency, wavelength and wavenumber
- understand and be able to use the Beer–Lambert law in calculation of sample absorbance, concentration or extinction coefficient
- understand the difference between specific absorbance and molar extinction coefficient, and be able to convert from one to the other
- be able to recognise chromophores within drug structures
- be able to use and construct UV calibration curves
- have an appreciation of the molecular basis of bathochromic and hypsochromic shifts in UV-visible spectra
- be able to read and interpret infrared spectra, and appreciate how and why these spectra can be used to confirm the identity of unknown samples
- be able to interpret ^{1}H NMR spectra and be able to assign the various peaks seen to specific proton sets within a molecule
- understand the basis of mass spectrometry and appreciate its value in furnishing the relative molecular mass of a test compound
- understand the underlying principles of chromatography with particular reference to TLC, HPLC and GC
- be familiar with volumetric and gravimetric analysis as quantification methods.

Electromagnetic spectrum

- *Electromagnetic radiation* travels through space as waves of defined *frequency* and *wavelength* (Figure 15.1). The complete spectrum of electromagnetic radiation spans from low-energy radiowaves through to high-energy X-rays and gamma rays (Figure 15.2).
- The wavelength (λ, metres) and frequency (υ, cycles/second or Hertz) of electromagnetic waves are related through the equation:

$$c = \upsilon.\lambda$$

where c is the speed of light (3×10^8 m.s^{-1}).

Figure 15.1
Electromagnetic wave.

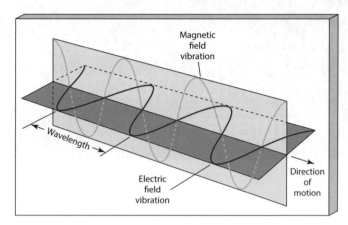

Figure 15.2
Electromagnetic
spectrum.

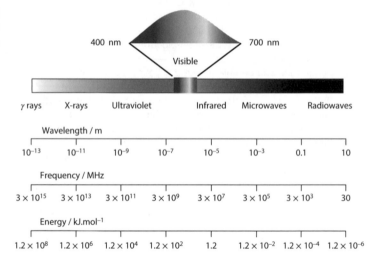

- The energy (E) associated with an electromagnetic wave is given by the equation:

$$E = h.\upsilon$$

where h is Planck's constant (6.63×10^{-34} $m^2.kg.s^{-1}$).

- When a molecule is exposed to electromagnetic radiation, the associated energy can be absorbed and cause:

1. an *electronic transition* (due to the promotion of an electron from a *bonding orbital* to a higher energy *antibonding orbital*)
2. a *vibrational transition* (associated with an increased vibration of atoms about a chemical bond)
3. a *rotational transition* (associated with a rotation of atoms about a chemical bond).

Rotational transitions require around 100 times more energy than electronic transitions, and these in turn require around 100 times more energy than vibrational transitions (Figure 15.2).

Electromagnetic wave

■ The region of the electromagnetic spectrum that the human eye can detect (which spans waves of wavelength 400–700 nm) is referred to as the *visible* spectrum.
■ Molecules that absorb electromagnetic radiation within the visible spectral region appear coloured. Molecules that absorb blue light will appear yellow or red in colour, while those that absorb red light will appear green or blue in colour.
■ The region of the electromagnetic spectrum that lies just beyond the red end of the visible spectrum is referred to as the *infrared* (*IR*) region.
■ The region of the electromagnetic spectrum that lies just beyond the violet end of the visible spectrum is referred to as the *ultraviolet* (or *UV*) region.
■ The interaction between molecules and electromagnetic radiation forms the basis of the subject of *spectroscopy*.

UV/visible spectroscopy

■ Quantitative spectroscopic analyses of drug substances most often involve *UV/visible spectroscopy*, exploiting their properties as regards absorption of UV and/or visible light.
■ Measurement of the amount of light absorbed by a sample is made in a UV/visible *spectrophotometer*. The extent of light absorption is governed by Beer's law and Lambert's law:
■ *Beer's law* states that the intensity of a beam of parallel, monochromatic light decreases exponentially with increasing concentration of the light-absorbing molecules. This is expressed mathematically as:

$$I = I_0.\exp(-\mathrm{k}.c)$$

where I and I_0 are respectively the intensity of the light incident on the sample and the intensity of the light after passing through the sample, c is the concentration of the light-absorbing species within the sample and k is a constant. Rearranging the equation above and taking logarithms gives:

$$\log_{10}(I_0/I) = \mathrm{k}.c$$

KeyPoints

■ Absorbed UV-visible light radiation causes electronic transitions in molecules.
■ Absorbed IR radiation causes vibrational transitions in molecules.

where $\log_{10}(I_0/I)$ is the (dimensionless) quantity referred to as *absorbance*.

- *Lambert's law* states that the intensity of a beam of parallel monochromatic light decreases exponentially as it travels through an increasing thickness of light-absorbing material. This is expressed mathematically as:

$$I = I_0.\exp(-k'.l)$$

where I and I_0 are as given above, l is the distance travelled by the light through the sample (the *path length*) and k' is a constant. Rearranging the equation given above and taking logarithms gives:

$$\log_{10}(I_0/I) = k'.l$$

- The Beer's law and Lambert's law equations are combined in the *Beer–Lambert law* equation:

$$\log_{10}(I_0/I) = k''.c.l$$

where k'' is a proportionality constant with units that depend on the units employed in specifying the concentration, c, and the path length, l.

- In UV/vis spectroscopy, the path length, l, is typically 1 cm but path lengths of 2 mm or even 1 mm are sometimes employed.
- If the concentration of the light-absorbing sample is measured in moles.L^{-1}, the proportionality constant in the Beer–Lambert equation is referred to as the *molar extinction coefficient*. It is given the symbol ε and has units of L.mol^{-1}.cm^{-1}.
- If the concentration of the light-absorbing sample is measured in g/100 mL (that is, in units of % w/v), the proportionality constant in the Beer–Lambert equation is referred to as *specific absorbance*. It is given the symbol A (1%, 1 cm) and has units of dL.g^{-1}.cm^{-1}.
- A (1%, 1 cm) and ε may be interconverted through the equation:

$$\varepsilon = A\ (1\%,\ 1\ cm)\ .\ (RMM/10)$$

where RMM is the relative molecular mass of the light-absorbing molecule.

- The chemical groups within a molecule that are responsible for its absorption of light are called *chromophores*.
- Chromophores may involve double bonds or triple bonds, and multiple bonds that are conjugated. The more extensive the conjugation within a chromophore, the greater its extinction coefficient, and the longer the wavelength of the absorbed light (Figure 15.3).
- For any given compound, a plot of its light absorbance as a function of wavelength is referred to as the absorption spectrum of the compound (Figure 15.3). The wavelength at which the

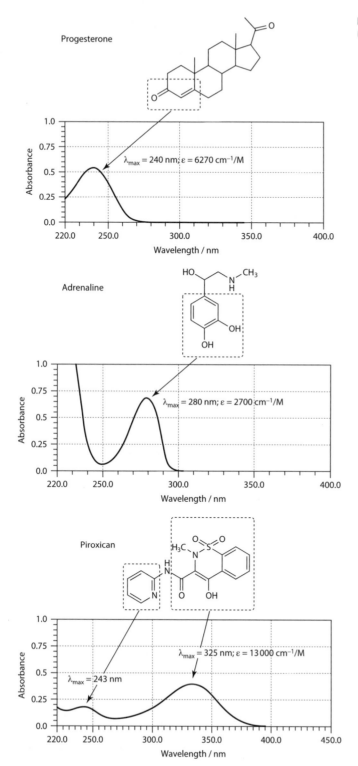

Figure 15.3
Ultraviolet chromophores.

KeyPoints

- Chromophores are the chemical groups within a molecule that are responsible for its absorption of UV-visible light
- Chromophores are chemical groups that involve multiple bonds, in particular conjugated bonds.

Tips

When using the Beer–Lambert law equation, ensure that you use the correct units:

- If you are using (or want to calculate) the molar extinction coefficient, path length must be in units of centimetres, and the concentration of the analyte must be in units of moles per litre.
- If you are using (or want to calculate) the specific absorbance, path length must be in units of centimetres, and the concentration of the analyte must be in units of grams per 100 millilitres.

absorbance is a maximum is referred to as λ_{max}, and the absorbance at this wavelength is referred to as A_{max}.

- Chromophores that have attached acidic or basic groups (e.g. -OH or -NH$_2$ groups) have UV spectra that are sensitive to pH. The ionisation/protonation of the acidic/basic group causes a change in the electronic structures of the chromophore, and so causes a shift in the molecular absorption spectrum.
- The UV absorbance of a compound is also often influenced by the nature of the solvent in which it is dissolved (Figure 15.4).
- If the excited state of a molecule is more polar than its ground state (as will be the case for π–π* transitions), the excited state will be stabilised to a greater extent in a polar solvent. The energy thus required for the electronic transition will be less, and the wavelength of light absorbed to effect the transition will be longer. The resulting shift in λ_{max} is referred to as a *bathochromic shift* or *red shift*.
- Conversely, if the ground state of a molecule is more polar than its excited state (as with n-π* transitions), it will be the ground state of the molecule that is stabilised more in the polar solvent. The energy required for the

Figure 15.4
Effects of pH on ultraviolet absorbance: paracetamol: bathochromic shift (lower panel); benzocaine: hypsochromic shift (upper panel).

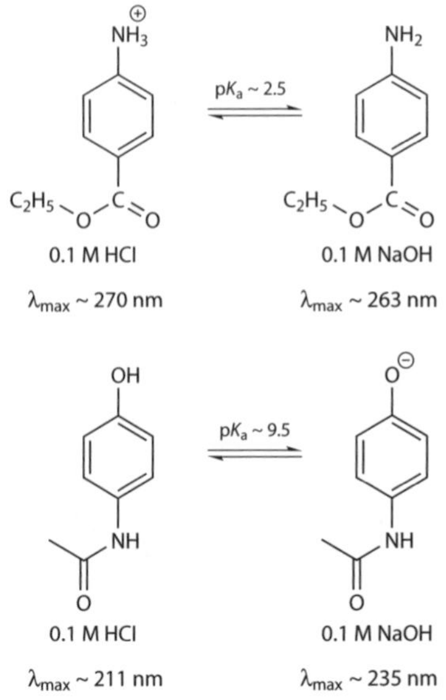

electronic transition will then be greater, and the wavelength of light absorbed to effect the transition will be shorter. The resulting shift in λ_{max} is referred to as a *hypsochromic shift* or *blue shift*.

- In aromatic amines, protonation of the amine group occupies the nitrogen lone pair electrons in (NH) bond formation, and they are no longer then conjugated with the π-electrons in the aromatic ring. The protonation of the amine group thus leads to a hypsochromic shift (Figure 15.4).
- In phenolic compounds, ionisation of the phenolic group (to give a phenoxide anion) results in the creation of a full negative charge on the oxygen. The associated electrons then interact more effectively with the π-electrons in the aromatic ring. The ionisation of the phenol group thus leads to a bathochromic shift (Figure 15.4).
- Spectroscopic drug assays often involve the use of UV *calibration curves* (Figure 15.5). Such curves are obtained by measuring the UV absorbance of solutions of the drug at different concentrations, and then plotting absorbance vs concentration to obtain (what is, hopefully) a straight line of positive slope that passes through the origin. The concentration of an unknown sample of the same drug can then be determined by following the same experimental procedure to measure its UV absorbance and extrapolating its concentration from the graph.

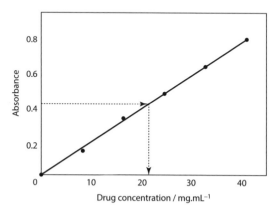

Figure 15.5
Ultraviolet calibration curve.

Infrared spectroscopy

- IR waves have wavelengths in the range 2.5–15 μm, but they are generally described with reference to their *wavenumber*, rather than their wavelength. Wavenumber is simply the reciprocal of wavelength, and (for IR waves) is typically expressed in units of cm^{-1}.

- IR spectra (Figure 15.6) are presented using the quantity *per cent transmittance*, rather than absorbance. This is given as:

$$\%T = 100.(I/I_0)$$

- IR spectra often show 20 or more peaks (these are generally referred to as *IR bands*), each of which can be assigned to a particular type of vibration of a particular chemical group.
- Single-bonded chemical groups, like O–H, N–H and C–H, absorb in the high-frequency IR region, 4000–2100 cm⁻¹. Triple-bonded groups, such as -CN, absorb around 2200–1500 cm⁻¹, while double-bonded groups, like C=O and C=C, absorb in the region of 1900–1500 cm⁻¹.
- Carbonyl groups give characteristically strong and sharp IR bands, and their position in the IR spectrum is often diagnostic for the type of carbonyl (e.g. ester, ketone, carboxylic acid: Table 15.1).
- The region of the IR spectrum at wavenumbers below 1500 cm⁻¹ is known as the *fingerprint region* (Figure 15.6). The IR bands in this region are not readily assigned to specific molecular vibrations, and are due to the stretching of a molecule as a whole.

Figure 15.6
Infrared spectrum of procaine.

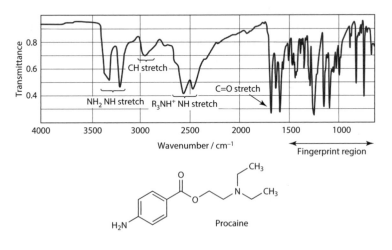

Procaine

Table 15.1 Carbonyl infrared bands

Functional group	Structure	C=O stretch/cm⁻¹
Aldehyde (saturated)	R-CHO	1730
Ketone (saturated)	RC(=O)R	1715
Ester	R-COOR	1745
Carboxylic acid	R-COOH	1715, 1580
Amide	R-CONH₂	1650
Anhydride	RC(=O)-O-C(=O)R	1815, 1765

Nuclear magnetic resonance spectroscopy

- Atomic nuclei that have unpaired protons or neutrons possess a property referred to as *magnetic spin*.
- When the nuclei with a magnetic spin are placed in an applied magnetic field, they behave like small bar magnets and align themselves in the direction of the applied field. If the aligned nuclei are then irradiated with radiowaves of the right frequency, they gain energy (moving from the ground state to the excited state) and become reoriented in opposition to the applied magnetic field (Figure 15.7). When the irradiation stops, the nuclei relax back to their ground state and in the process release energy of radiowave frequency.
- The *nuclear magnetic resonance* (*NMR*) *spectrometer* is an instrument designed to exploit this *magnetic resonance* phenomenon.
- Those nuclei that have a magnetic spin of ½ are those most commonly used in *NMR spectroscopy*. Such nuclei include ^1H (proton), ^{13}C (carbon-13), ^{15}N (nitrogen-15) and ^{31}P (phosphorus-31).
- ^1H NMR requires relatively little material and is fairly quick to perform. ^{13}C NMR requires significantly larger amounts of material and is much slower to perform.
- The frequency of radiowave required to flip an atomic nucleus within an applied magnetic field is exquisitely sensitive to the electronic environment of the nucleus. The magnetic resonance of

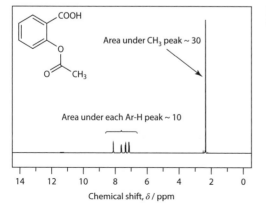

Figure 15.7
^1H NMR spectrum of aspirin.

the protons within a CH_3 group at the end of an alkyl chain is thus different from that of the protons within a CH_3 group attached to a carbonyl carbon, and both of these resonances are very different from that of the protons attached to an aromatic ring. NMR thus provides an excellent means by which to determine what kinds of chemical groups are present in a molecule, and it provides a very good way, therefore, of elucidating the chemical structure of an unknown compound.

■ The resonance frequency of a given atomic nucleus will vary as a function of the power/applied magnetic field strength of the magnet in the NMR spectrometer. Typical field strengths are 300, 400 and 600 MHz.

■ Even in spectrometers with the same design and the same magnet, however, the magnetic field strength will vary somewhat, and this will mean that a given nucleus in a given compound will exhibit different resonance frequencies when measured in different spectrometers.

■ For this reason, the resonance frequency of a nucleus is normalised to that of a nucleus in a reference compound:

$$\delta = \frac{\text{frequency (Hz)} - \text{frequency TMS (Hz)}}{\text{operating frequency (MHz)}}$$

■ The normalised resonance frequency, δ, is known as the *chemical shift* of the nucleus and has units of *ppm* (parts per million).

■ In ¹H-NMR spectroscopy, the reference compound, also referred to as an *internal standard*, is trimethylsilane (TMS). The reference resonance frequency is thus that of the 12 identical ¹H nuclei in the $Si(CH_3)_4$ molecule.

■ Some typical values of proton NMR chemical shifts are given in Table 15.2.

■ Electrons circulating around protons placed in a magnetic field (in the NMR spectrometer) will set up a local magnetic field that opposes that of the applied magnetic field. The protons are then

Table 15.2 ¹H NMR chemical shifts

Proton	Approximate chemical shift/ppm
-CH₃	0.9
-CH₂-	1.3
-CH-	1.5
-CH=CH-	5–6
Ar-H	7
R-CHO	9–10
Ar-OH	4–7
R-COOH	10–12
R-NH₂	1.5–4
R-OH	2–5

said to be *shielded*, and their NMR peaks appear *upfield* (to the right in the NMR spectrum).

- Protons that are adjacent to electronegative groups, however, give rise to ^{1}H NMR signals that appear *downfield* (that is, towards the left of the NMR spectrum). The electronegative groups draw electrons away from the adjacent protons, and the protons are then said to be *deshielded*.

- Integration of the areas under the peaks in a ^{1}H NMR spectrum gives an indication of the relative numbers of protons involved. So, for example, in the ^{1}H NMR spectrum for aspirin (Figure 15.7), the signal at $\delta = 2.35$ ppm (which is due to the methyl protons) has three times the area of the peaks seen around $\delta = 7$–8 ppm (which are due to the protons on the aromatic ring). This is because the methyl group involves *three* (*chemically equivalent*) protons, while each *one* of the four protons on the aromatic ring is chemically non-equivalent and gives rise to a separate NMR signal.

- As the result of a phenomenon known as *spin-spin coupling*, ^{1}H NMR signals may sometimes be split to give multiple peaks, with the multiplicity depending on the number of equivalent protons on the adjacent atom(s). If there are n equivalent protons on the adjacent carbon, the ^{1}H NMR signal is split into $n + 1$ peaks. An NMR peak split into two peaks is said to form a *doublet*; if it is split into three or four peaks, it is said to form a *triplet* or *quartet*.

- In the ^{1}H NMR spectrum for ethanol, therefore (Figure 15.8), the peak (at $\delta \sim 1.3$ ppm) that arises due to resonance of the three methyl protons is split into three peaks with intensities in the ratio 1:2:1. These three peaks are caused by the two protons bonded to the adjacent CH_2 group. In analogous manner, the peak (at $\delta \sim 3.6$ ppm) that arises due to resonance of the CH_2 protons

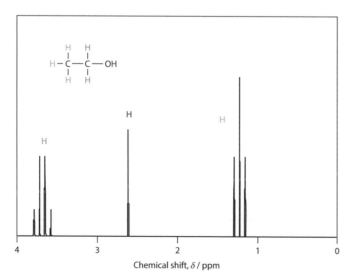

Figure 15.8
^{1}H nuclear magnetic resonance spectrum of ethanol.

Chemical shift, δ / ppm

KeyPoints

- The position of a peak in a ^{1}H-NMR spectrum is determined by the electronic environment of the proton(s).
- The multiplicity/splitting of a peak in a ^{1}H-NMR spectrum is determined by the number of protons on the attached carbon atom(s). The peak due to protons that have n protons on an attached carbon atom will be split into a multiplet of $n + 1$ peaks.
- If D_2O is added to an NMR sample, the ^{1}H peaks due to acidic, exchangeable protons (on OH, NH and SH groups) will disappear.

is split into four peaks (with relative intensities in the ratio 1:3:3:1) by the three protons in the methyl group.

- The addition of a few drops of D_2O to an NMR sample can be helpful in identifying the ^{1}H resonances due to -OH, -NH$_2$ and -SH protons. The weakly acidic protons in these chemical groups are susceptible to *deuterium exchange.* That is, they can exchange with deuterons from the D_2O. When they do so, the resulting -OD, -ND$_2$ and -SD groups do not give rise to peaks in the ^{1}H NMR spectrum. By comparing the NMR spectrum before and after the addition of D_2O, therefore, and noting which peaks disappear when the D_2O is added, the labile proton resonances are identified.

Mass spectrometry

- Although *mass spectrometry* is *not* a spectroscopic technique, it is included here because it provides valuable information of use in elucidating the structure of an unknown compound.
- In this technique, a sample of the material under study is subjected to *chemical ionisation* or *electron ionisation* and the resulting charged species are then accelerated into a vacuum chamber and separated on the basis of their *mass to charge ratio* (*m/z ratio*: Figure 15.9).
- The charged species produced in the mass spectrometer will generally include a range of different fragment ions and, under the right conditions, will also include the *molecular ion,* which is simply the intact test molecule minus one electron (a *molecular ion radical*).
- The peak due to the molecular ion (the *M$^+$ peak*) will usually be the peak with the highest *m/z*. Identification of the M$^+$ peak for a test compound will afford the relative molecular mass of the compound.

Figure 15.9
Mass spectrum of ibuprofen (relative molecular mass, 206).

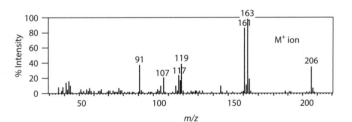

Chromatography

- Chromatography is the separation of the individual components of a mixture by exploiting differences in how these components partition between a mobile phase and a stationary phase.
- A sample mixture is dissolved in a mobile phase, which can be either a gas or a liquid. The mobile phase is then passed through an immobile stationary phase.
- High-performance liquid chromatography (HPLC) and gas chromatography (GC) use columns which are tightly packed with stationary phase and through which the mobile phase is passed (Figure 15.10). The continuous addition of mobile phase transports the components of the sample through the stationary phase in a process termed elution.
- Components that interact strongly with the stationary phase spend more time in the column and are separated from those components that only interact weakly with the stationary phase and prefer to remain in the mobile phase. These components will move through the column faster and as a result elute first.
- As the individual components elute from the column they can be quantified by a detector or collected for further analysis.

> **KeyPoint**
> - The individual components of a sample will interact differently with the mobile phase and the stationary phase.

> **KeyPoint**
> - If component B interacts weakly with the stationary phase compared with component A, component B will elute before component A (Figure 15.10).

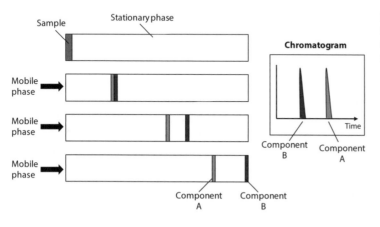

Figure 15.10
Separation of a mixture by chromatography.

Thin-layer chromatography (TLC)

- TLC is a simple and rapid chromatographic method that can be used to monitor the progress of a reaction or assess the purity of a compound.

Figure 15.11
Thin-layer
chromatography (TLC).

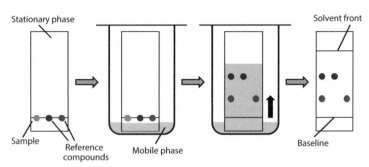

- TLC involves a stationary phase immobilised on a glass, aluminium or plastic plate, and an organic solvent as the mobile phase (Figure 15.11).
- The sample, either a liquid or a solid dissolved in a volatile solvent, is deposited as a spot on the stationary phase.
- The individual components of a sample can be identified by simultaneously running reference compounds alongside the sample on the same plate.
- The bottom edge of the plate is placed in a solvent reservoir and the solvent moves up the plate by capillary action.
- TLC separates compounds based on their polarity.
- If a silica plate is used in combination with a non-polar solvent, the polar components of a mixture will interact with the silica more strongly than the non-polar components. Non-polar components will move up the silica plate faster than the polar components as they are only interacting weakly with the stationary phase.
- Increasing the polarity of the mobile phase will reduce the difference in polarity between the two phases. As a result, polar components will not interact as strongly with the stationary phase and will move more rapidly up the silica plate compared with using a less polar mobile phase.
- When the solvent front reaches three-quarters of the height of the stationary phase, the plate is removed from the solvent reservoir. The separated spots can then be visualised with UV light or by staining with a suitable reagent (e.g. ninhydrin for amino acids).
- Individual components within a mixture can be identified by comparison with the distance travelled of the reference compounds.

Tips

- The reservoir must be sealed to create a solvent-saturated atmosphere, preventing the evaporation of mobile phase from the plate.
- Placing filter paper in the solvent at the back of the tank will further aid in creating a saturated atmosphere.
- Mark the height of the solvent front immediately after removing the plate from the solvent reservoir.
- Always measure from the centre of a component spot.

- Retention factor,

$$R_f = \frac{\text{Distance travelled by the component above the baseline}}{\text{Distance travelled by the solvent front above the baseline}}$$

High-performance liquid chromatography

- HPLC can also be used to monitor the progress of a reaction and assess the purity of a compound, but, unlike TLC, HPLC can be used to quantify the components of a mixture.
- HPLC consists of a stationary phase of tightly packed particles encased in a steel column, and a solvent as the mobile phase (Figure 15.12).
- The components of a mixture are separated by injecting the sample on to the column. The different components of the mixture pass through the column at different rates due to differences in the way the components interact with the mobile phase and the stationary phase.
- High-pressure pumps are used to increase the efficiency of the separation of individual components of a mixture.
- The components are detected by monitoring the mobile phase as it leaves the column for changes in UV-visible absorption, fluorescence or refractive index.
- The chromatogram obtained provides the retention time for each component of the mixture. The relative quantities of the components present in the original sample are proportional to the areas of the individual peaks in the chromatogram.
- The concentration of an individual component within the mixture can be established by constructing a calibration

KeyPoints

- High-performance liquid chromatography can operate in two different modes: normal phase and reverse phase.
- Normal-phase chromatography utilises a polar stationary phase and a non-polar mobile phase.
- Reverse-phase chromatography utilises a non-polar stationary phase and a polar mobile phase (usually water).
- Whereas polar components elute last in normal-phase chromatography, they will elute first in reverse-phase chromatography.

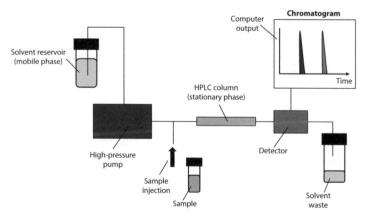

Figure 15.12
High-performance liquid chromatography (HPLC).

graph using chromatograms obtained for known concentrations of the component to be quantified.

Gas chromatography

- GC can also be used to monitor the progress of a reaction and assess the purity of a compound and, as with HPLC, can be used to quantify the components of a mixture. However, GC can only be applied to volatile organic compounds.
- GC consists of a stationary phase of a high-boiling liquid coated on to the surface of a solid support and an inert gas (nitrogen, helium or argon) as the mobile phase (Figure 15.13).
- The components of a mixture are separated by injecting the sample on to the column which is located in a heated oven. This vaporises the sample and the individual components of the mixture pass through the column at different rates due to differences in the way the components interact with the mobile phase and the stationary phase.
- Mixtures containing components with significantly different boiling points can be separated by starting at a low oven temperature and increasing the temperature over time to elute the higher-boiling components.

Figure 15.13
Gas chromatography (GC).

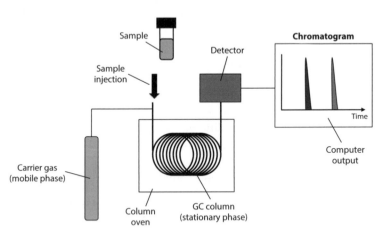

Volumetric (titrimetric) analysis

- Volumetric analysis is a quantitative analytical technique in which a reaction occurs between a solution of known concentration and a solution of unknown concentration.
- A titration method is used in which a solution containing a known concentration of one reactant (the standard solution) is added to a solution containing an unknown concentration of a second reactant.
- The equivalence point is reached when all of the second reactant has been consumed. The number of molecules of the reactant of

known concentration now equals the number of molecules of the reactant of unknown concentration.

■ Where an indicator is used to show that all of the reactant of unknown concentration has been consumed, the onset of the colour change is called the endpoint. The equivalence and endpoints will not be the same since at the endpoint there will usually be a slight excess of the reactant that is being added which is giving the observed colour change.

■ Volumetric analysis methods make use of acid/base reactions, precipitation reactions, complexation reactions and redox reactions.

Gravimetric analysis

■ Gravimetric analysis is a quantitative analytical technique that is based on the isolation of a substance by precipitation and the weighing of this precipitate.

■ A weighed sample will be dissolved to give a solution to which a precipitating reagent is added. The precipitating reagent will cause one component of a mixture to precipitate selectively from the solution. The precipitate will be filtered, dried and weighed; the content of the component can then be calculated.

■ The component to be analysed must be completely precipitated from solution, hence an excess of the precipitating reagent is used.

■ It is essential that the precipitate is a pure substance of known composition with no impurities.

Self-assessment

1. **Which of the drug molecules shown below would you expect to give the strongest absorbance of UV/visible light?**

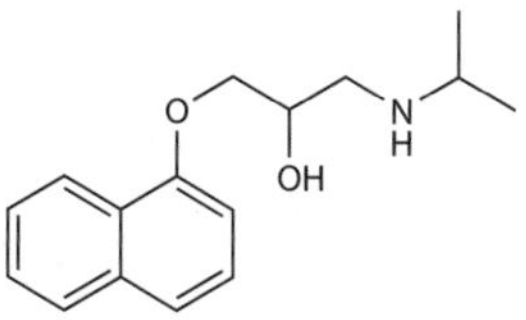

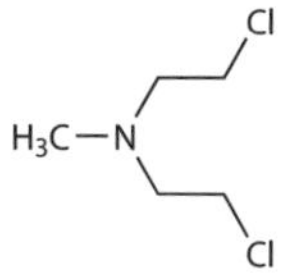

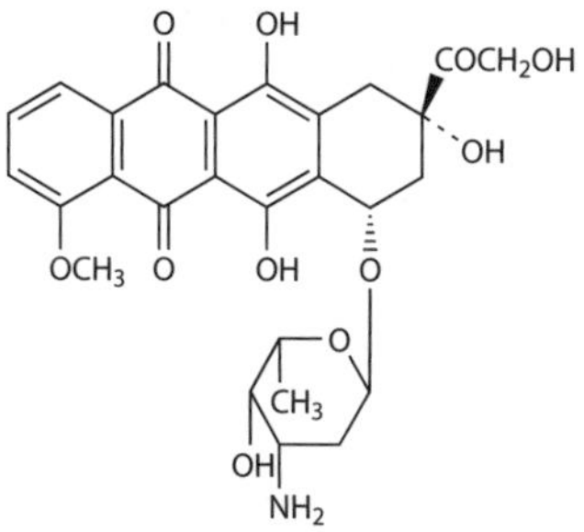

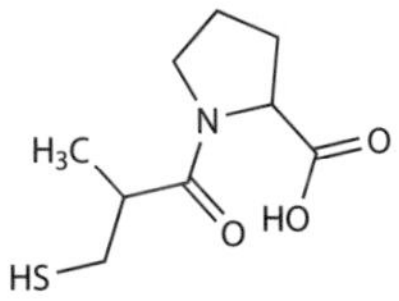

a. drug A
b. drug B
c. drug C
d. drug D

2. Given that the specific absorbance, A (1%, 1 cm) of aceclofenac (in methanol solution, at 275 nm) is 330 dL.g^{-1}.cm^{-1}, what will be the concentration of the drug present in a sample with an absorbance of 0.33 (measured at 275 nm, in a 2 mm path length cell)?

a. 0.005 g.mL^{-1}
b. 0.005 % w/v
c. 0.001 g.mL^{-1}
d. 0.001 % w/v

3. A drug absorbs infrared radiation at wavenumber 1760 cm^{-1}, giving an absorbance of 0.5. This corresponds to a percentage transmittance of:

a. 0.316
b. 3.16
c. 31.6
d. 316

4. Given the UV calibration curve for solutions of aspirin shown below:

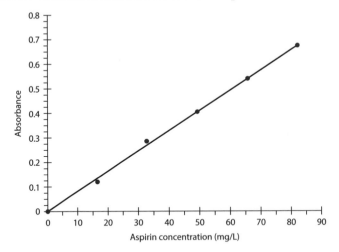

estimate the concentration of the drug contained in a sample that gives a measured absorbance (under identical experimental conditions) of 0.6:

a. 0.07 g per 100 mL
b. 0.007 % w/v
c. 70 % w/v
d. 70 mg.mL^{-1}

5. The fingerprint region in an infrared spectrum covers the range:
a. $4000–2750\ cm^{-1}$
b. $2750–2000\ cm^{-1}$
c. $2000–1500\ cm^{-1}$
d. $1500–750\ cm^{-1}$

6. Electromagnetic radiation with a wavelength of 5 µm will have a wavenumber of:
a. $2 \times 10^5\ cm^{-1}$
b. $2 \times 10^3\ cm^{-1}$
c. $2 \times 10^{-5}\ cm^{-1}$
d. $2 \times 10^{-3}\ cm^{-1}$

7. The carbonyl stretch of a saturated ester group gives rise to an IR band around:
a. $750\ cm^{-1}$
b. $1750\ cm^{-1}$
c. $2500\ cm^{-1}$
d. $3500\ cm^{-1}$

8. Magnetic resonance of the CH_2 protons in benzocaine (shown below) gives rise to:

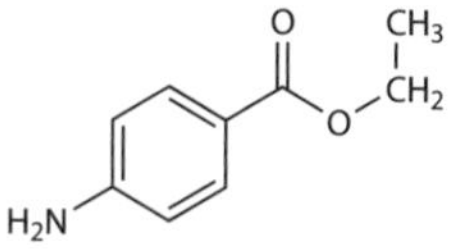

a.   a doublet at 3.5 ppm
b.   a doublet at 1 ppm
c.   a quartet at 3.5 ppm
d.   a quartet at 1 ppm

9. Ignoring any multiplicity arising due to spin–spin coupling, how many peaks do you expect to see in the ^1H NMR spectrum of *t*-butylamine?
a.   10
b.   5
c.   4
d.   2

10. Which of the following techniques would furnish the greatest amount of information to be used in determining the structure of an unknown compound?
a.   UV spectroscopy
b.   mass spectrometry
c. ^1H NMR
d.   IR spectroscopy

11. **Which of the following statements is true about HPLC?**
a. The mobile phase must be water.
b. The column is normally constructed from stainless steel to overcome the mechanical stress involved with the flowing mobile phase.
c. The detector must be located before the column.
d. The column is normally only 1 cm long.

12. **Which of the following statements is true about TLC?**
a. The thin layer coats the inside of a specially designed column.
b. The most polar components elute fastest.
c. The mobile phase is pumped through the system at high pressure.
d. R_f is an abbreviation for retention factor.

chapter 16
Drug licensing and pharmacopoeia

Overview

After learning the material presented in this chapter you should:

- appreciate the role of the UK Medicines and Healthcare products Regulatory Agency, and know how medicinal products are licensed for use in the UK
- know the essential information that must be provided for pharmaceutical companies to obtain marketing authority for new medicinal products
- know the types of information that are provided on drugs and medicinal products in the monographs of the *British Pharmacopoeia* (the BP)
- understand the nature and purpose of identification tests given in BP monographs
- understand the nature and purpose of limit tests given in BP monographs
- appreciate the role of the National Institute for Health and Care Excellence.

Drug licensing

The UK government (via the Department of Health) requires that all medicines and medical devices are licensed and safe for use. The task of regulation and the approval of licences for drugs and medicines are carried out by the *Medicines and Healthcare products Regulatory Agency* (*MHRA*). The range of products regulated by the MHRA includes: medicines, medical devices, materials derived through tissue engineering (e.g. stem cells), surgical implants, prostheses and blood-derived products (e.g. clotting agents).

The MHRA is organised as a number of divisions, including: *Inspection and Standards* (which is concerned with ensuring that the required standards are met in the manufacture and supply of medicines and medical devices in the UK), *Licensing* (which has the remit to receive and approve applications for marketing authorisation for new medicines, new formulations and new devices), and *Vigilance and Risk Management of Medicines* (which is concerned with safeguarding public health through the promotion of safe use of medicines).

KeyPoints

- In the UK, the MHRA is charged with the task of ensuring that all medicines and medical devices are licensed and safe for use.
- The MHRA is the body that provides marketing authorisation for new drug products.

Marketing authorisations

Marketing authorisation for new drug products requires that the manufacturing company provides comprehensive details to (the Licensing Division of) the MHRA, setting out evidence to show that manufacture of the product is appropriately controlled and validated, and yields material of an acceptable quality.

The submission document details:

- the methodology involved in preparation of the active agent
- spectroscopic information to confirm its identity (and also the identities of any byproducts)
- the results of stability testing and the associated product degradation rate and calculated shelf-life
- the details of the product's formulation
- clinical data obtained through studies using both animals and humans
- details of the product's packaging, labelling and instructions for use.

British Pharmacopoeia

Authoritative official standards for all pharmaceutical substances and medicinal products are provided by the multivolume reference text, the *British Pharmacopoeia* (frequently referred to simply as the *BP*). The equivalent reference work for EU countries is the *European Pharmacopoeia*.

The BP comprises a series of *monographs* for medicinal and pharmaceutical substances, together with general and specific information on their *formulated preparations*, information on *herbal medicinal products,* the materials used in manufacture of *homeopathic preparations*, information on *blood*, *radiopharmaceutical* and *immunological products* and *infrared reference spectra*. It also includes a volume dealing specifically with *veterinary products*, and a large series of *Appendices* that provide information on pharmaceutical analytical techniques, with specifications given for the various reference materials needed, the apparatus and equipment required and the protocols to be followed in the performance of the various analyses and tests. In addition, there is a listing of the *British Approved Names* for all medicinal and pharmaceutical substances licensed for use in the UK.

KeyPoints

- The *British Pharmacopoeia* defines the standards required to ensure the quality of medicines, with the aim of protecting public health.

Drug monographs

Each of the *product-specific BP monographs* will provide details of: the chemical structure of the drug product (if applicable/known), a description of its physical characteristics (colour, solubility, melting point, etc.), identification tests for the drug (including chemical/colour tests and reference spectroscopic data) and limit tests for likely impurities.

The BP additionally provides *General Monographs,* which contain details that apply to given classes of medicine (relating, therefore, to tablets, dispersible tablets and suspensions).

Identification tests

The BP monograph for a given drug product will specify tests that can be performed to confirm its identity.

First identification tests are those that can be performed to confirm identity in all circumstances.

Second identification tests are those that can be used for identification in cases where the substance or preparation to be tested is fully traceable to a batch that has been certified to comply with all other requirements of the monograph.

The identification tests given in the BP monographs are *not* designed to provide full confirmation of the chemical structure or composition of a pharmaceutical/medicinal product. Rather, they are intended to give confirmation, with an acceptable degree of assurance, that the product conforms to the description given on the label.

Limit tests

Each BP monograph gives details of *limit tests* that can be used to quantify the levels of impurities in a given drug product, either arising as byproducts of the chemical synthesis or extraction and purification process(es) employed in preparation of the drug, or else through decomposition of the drug on storage. Other common limit tests include tests for *heavy metals* (as a means of protecting against the associated heavy metal toxicities), *sulfated ash* (as a means of judging the levels of inorganic impurities), *moisture content* (to ensure that products required dry are provided dry) and *microbial content* (to protect against microbial contamination in products that are required to be sterile).

Using a drug monograph

KeyPoint

- BP drug monographs provide details of the physicochemical characteristics of the drug, together with descriptions of identification tests and limit tests (for impurities).

British Pharmacopoeia Volumes I and II
Monographs: Medicinal and Pharmaceutical Substances

Aspirin
General Notices

- Hyperlinks to relevant sections of BP, including appendices

(Acetylsalicylic Acid, Ph Eur monograph 0309)

- Name and monograph reference number

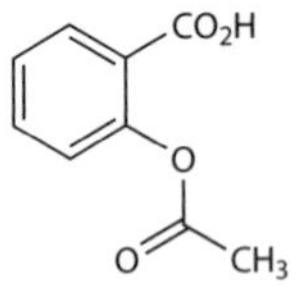

$C_9H_8O_4$ 180.2   *50-78-2*

- Chemical structure

- Molecular formula, molecular weight and Chemical Abstracts Service (CAS) registry number

Action and use
Salicylate; non-selective cyclo-oxygenase inhibitor; antipyretic; analgesic; anti-inflammatory.

- Class of drug, mode of action and therapeutic use

Preparations
Aspirin Tablets
Dispersible Aspirin Tablets
Effervescent Soluble Aspirin Tablets
Gastro-resistant Aspirin Tablets
Aspirin and Caffeine Tablets
Co-codaprin Tablets
Dispersible Co-codaprin Tablets

- Drug formulations containing aspirin

Ph Eur

DEFINITION
2-(Acetyloxy)benzoic acid.

- IUPAC systematic name

Content
99.5 per cent to 101.0 per cent (dried substance).

- Limits of content, as determined using the assay described in the monograph

CHARACTERS

- Physical properties of aspirin: appearance, solubility and melting point

Appearance
White or almost white, crystalline powder or colourless crystals.

Solubility
Slightly soluble in water, freely soluble in ethanol (96 per cent).

mp
About 143 °C (instantaneous method).

IDENTIFICATION
First identification A, B.
Second identification B, C, D.
A. Infrared absorption spectrophotometry *(2.2.24)*.
   Comparison acetylsalicylic acid CRS.
B. To 0.2 g add 4 mL of *dilute sodium hydroxide solution R* and boil for 3 min. Cool and add 5 mL of *dilute sulfuric acid R*. A crystalline precipitate is formed. Filter, wash the precipitate and dry at 100-105 °C. The melting point *(2.2.14)* is 156 °C to 161 °C.
C. In a test tube mix 0.1 g with 0.5 g of *calcium hydroxide R*. Heat the mixture and expose to the fumes produced a piece of filter paper impregnated with 0.05 mL of *nitrobenzaldehyde solution R*. A greenish-blue or greenish-yellow colour develops on the paper. Moisten the paper with *dilute hydrochloric acid R*. The colour becomes blue.
D. Dissolve with heating about 20 mg of the precipitate obtained in identification test B in 10 mL of *water R* and cool. The solution gives reaction (a) of salicylates *(2.3.1)*.

Identification test A

- Identification of key functional groups
- Comparison of fingerprint region with the authentic compound

Identification test B

- Ester hydrolysis under basic conditions followed by protonation in the second step
- Product is salicylic acid as confirmed by the melting point of the product

Identification test C

- Acetate is converted to acetone in the first step (mechanism beyond the scope of this book).
- Initial aldol condensation of acetone with nitrobenzaldehyde which subsequently goes on to yield indigo (mechanism beyond the scope of this book).

Identification test D

- Addition of $FeCl_3$ forms a purple complex with salicylic acid.
- Also used as a test of aspirin purity, only phenol impurities will complex with $FeCl_3$ to give a purple solution.

TESTS

- Full details for sample preparation of substance to be tested and standards for comparison

Appearance of solution

The solution is clear (2.2.1) and colourless (2.2.2, Method II). Dissolve 1.0 g in 9 mL of *ethanol (96 per cent) R*.

Related substances

Liquid chromatography (2.2.29). *Prepare the solutions immediately before use.*

Test solution Dissolve 0.100 g of the substance to be examined in *acetonitrile for chromatography R* and dilute to 10.0 mL with the same solvent.

Reference solution (a) Dissolve 50.0 mg of *salicylic acid R* (impurity C) in the mobile phase and dilute to 50.0 mL with the mobile phase. Dilute 1.0 mL of the solution to 100.0 mL with the mobile phase.

Reference solution (b) Dissolve 10 mg of *salicylic acid R* (impurity C) in the mobile phase and dilute to 10.0 mL with the mobile phase. To 1.0 mL of the solution add 0.2 mL of the test solution and dilute to 100.0 mL with the mobile phase.

Reference solution (c) Dissolve with the aid of ultrasound the contents of a vial of *acetylsalicylic acid for peak identification CRS* (containing impurities A, B, D, E and F) in 1.0 mL of *acetonitrile R*.

Column:

■ High-performance liquid chchromatography conditions to be used

– *size:* l = 0.25 m, Ø = 4.6 mm;
– *stationary phase*: *octadecylsilyl silica gel for chromatography R* (5 µm).
Mobile phase *phosphoric acid R, acetonitrile for chromatography R, water R* (2:400:600 *V/V/V*).
Flow rate 1 mL/min.
Detection Spectrophotometer at 237 nm.
Injection 10 µL.
Run time 7 times the retention time of acetylsalicylic acid.

■ Comparison of chromatograms to identify impurities present

Identification of impurities Use the chromatogram obtained with reference solution (a) to identify the peak due to impurity C; use the chromatogram supplied with *acetylsalicylic acid for peak identification CRS* and the chromatogram obtained with reference solution (c) to identify the peaks due to impurities A, B, D, E and F.
Relative retention With reference to acetylsalicylic acid (retention time = about 5 min): impurity A = about 0.7; impurity B = about 0.8; impurity C = about 1.3; impurity D = about 2.3; impurity E = about 3.2; impurity F = about 6.0.

■ Details for identifying specific impurities

System suitability Reference solution (b):
– *resolution*: minimum 6.0 between the peaks due to acetylsalicylic acid and impurity C.

Limits:
– *impurities A, B, C, D, E, F*: for each impurity, not more than 1.5 times the area of the principal peak in the chromatogram obtained with reference solution (a) (0.15 per cent);
– *unspecified impurities*: for each impurity, not more than 0.5 times the area of the principal peak in the chromatogram obtained with reference solution (a) (0.05 per cent);
– *total*: not more than 2.5 times the area of the principal peak in the chromatogram obtained with reference solution (a) (0.25 per cent);

– *disregard limit*: 0.3 times the area of the principal peak in the chromatogram obtained with reference solution (a) (0.03 per cent).

- Limits for impurities based on HPLC comparison with reference solution
- The aspirin sample should lie within these limits

- Specific limits for heavy-metal content, loss on drying and sulfated ash
- Full experimental details for these tests are described in the appropriate sections of the BP

Heavy metals *(2.4.8)*
Maximum 20 ppm.
Dissolve 1.0 g in 12 mL of *acetone R* and dilute to 20 mL with *water R*. 12 mL of the solution complies with test B. Prepare the reference solution using lead standard solution (1 ppm Pb) obtained by diluting *lead standard solution (100 ppm Pb) R* with a mixture of 6 volumes of *water R* and 9 volumes of *acetone R*.

Loss on drying *(2.2.32)*
Maximum 0.5 per cent, determined on 1.000 g by drying *in vacuo*.

Sulfated ash *(2.4.14)*
Maximum 0.1 per cent, determined on 1.0 g.

- Acid/base titration to determine acid content
- Solution initially pink and turns clear at endpoint

ASSAY
In a flask with a ground-glass stopper, dissolve 1.000 g in 10 mL of *ethanol (96 per cent) R*. Add 50.0 mL of *0.5 M sodium hydroxide*. Close the flask and allow to stand for 1 h. Using 0.2 mL of *phenolphthalein solution R* as indicator, titrate with *0.5 M hydrochloric acid*. Carry out a blank titration.
1 mL of *0.5 M sodium hydroxide* is equivalent to 45.04 mg of $C_9H_8O_4$.

STORAGE
In an airtight container.

- Storage conditions for aspirin

IMPURITIES
Specified impurities A, B, C, D, E, F.

■ Chemical structures of possible impurities

■ A and B would be present from using impure starting materials during manufacture
■ C could be present from manufacture or hydrolysis of aspirin after synthesis
■ D and E are dimeric compounds originating from the synthesis, where esterification occurs between two salicylic acid molecules
■ F is also a dimeric compound originating from the synthesis. Here the carboxylic acid groups of two salicylic acid molecules are linked as an anhydride.

Health care standards

Guidance and advice on medicines' management, optimisation and use are provided to NHS practitioners in England and Wales by the *National Institute for Health and Care Excellence* (*NICE*) – formerly known as the *National Institute for Clinical Excellence*.

NICE is an independent, non-departmental public body whose role is to improve outcomes for people using the NHS and other public health and social care services. It does so by:

■ producing evidence-based guidance and advice for health, public health and social care practitioners
■ developing quality standards and performance metrics for those providing and commissioning health, public health and social care services
■ providing a range of information services for commissioners, practitioners and managers across the spectrum of health and social care.

The many functions performed by NICE include the provision of:

- summaries of the best available evidence for selected new medicines, and for existing medicines with new indications, to inform local NHS planning and decision making
- summaries of the best available evidence on selected unlicensed and off-label medicines, designed to meet the demand for information required in local NHS planning and decision making
- good practice guidance for those in health and social care involved in handling, prescribing, commissioning and making decisions about medicines.

Self-assessment

1. **The BP monographs are prepared and published by the *British Pharmacopoeia* Commission, which is overseen by the:**
a. National Institute for Clinical Excellence
b. General Pharmaceutical Council
c. Medicines and Healthcare products Regulatory Agency
d. National Institute for Health and Care Excellence

2. **A BP monograph will *not* provide information on:**
a. drug metabolites
b. drug solubilities
c. the chemical structures of drugs
d. the available formulations of drugs

3. **The collected volumes and appendices of the *British Pharmacopoeia* do *not* give information on:**
a. the synthetic routes for drug substances
b. veterinary drug products
c. pharmaceutical excipients
d. infrared spectra of drug substances

4. **A white powder is suspected to be aspirin, and as a means of determining whether this is true or not, a pharmaceutical analyst might consult the relevant BP monograph and perform:**
a. the first identification test for acetylsalicylic acid
b. the second identification test for acetylsalicylic acid
c. assays to determine the levels of salicylic acid impurity
d. the limit tests described for sulfated ash and heavy metals

5. **In the UK, marketing authorisation for new drug products is granted by:**
a. the Secretary of State for Health
b. the Inspection and Standards division of the MHRA
c. the Licensing Division of the MHRA
d. the Chief Pharmaceutical Officer in the Department of Health

Answers to Self-assessment

Chapter 1: Chemical structure and bonding

1. b.
2. c.
3. a.
4. d.
5. b.
6. a.
7. b.
8. c.
9. d.
10. c.

Chapter 2: Intermolecular interactions

1. a.
2. c.
3. b.
4. b.
5. c.
6. d.
7. b.
8. c.
9. b.
10. a.

Chapter 3: Acids and bases

1. a.
2. b.
3. d.
4. c.
5. c.
6. c.
7. d.
8. a.
9. a.
10. c.

Chapter 4: Stereochemistry

1. c.
2. a.
3. d.
4. c.
5. a.
6. d.
7. b.
8. b.
9. d.
10. c.

Chapter 5: Chemical reaction mechanisms

1. b.
2. a.
3. c.
4. a.
5. d.
6. d.
7. c.
8. c.
9. d.
10. a.

Chapter 6: Chemistry of electrophiles and nucleophiles

1. a.
2. c.
3. c.
4. a.
5. d.
6. b.
7. a.
8. d.
9. a.
10. b.

Chapter 7: Chemistry of aromatic compounds

1. b.
2. a.
3. a.
4. c.

5. b.
6. d.
7. c.
8. a.
9. c.
10. c.

Chapter 8: Chemistry of carbonyl compounds

1. d.
2. c.
3. a.
4. a.
5. c.
6. b.
7. b.
8. c.
9. d.
10. a.

Chapter 9: Chemistry of aromatic heterocyclic compounds

1. b.
2. c.
3. b.
4. b.
5. c.
6. a.
7. d.
8. a.
9. c.
10. d.

Chapter 10: Amino acids, peptides and proteins

1. c.
2. b.
3. a.
4. b.
5. a.
6. b.
7. d.
8. d.
9. c.
10. d.

Chapter 11: Carbohydrates and nucleic acids

1. b.
2. c.
3. c.
4. d.
5. a.
6. c.
7. a.
8. a.
9. d.
10. a.

Chapter 12: Lipids and steroids

1. b.
2. b.
3. d.
4. c.
5. c.
6. b.
7. d.
8. a.
9. c.
10. d.

Chapter 13: Chemical stability of drugs

1. c.
2. b.
3. a.
4. d.
5. a.
6. d.
7. c.
8. a.
9. b.
10. d.

Chapter 14: Drug metabolism

1. c.
2. c.
3. d.
4. b.
5. c.

6. d.
7. c.
8. a.
9. c.
10. c.

Chapter 15: Molecular spectroscopy and pharmaceutical analysis

1. c.
2. b.
3. c.
4. b.
5. d.
6. b.
7. b.
8. c.
9. d.
10. c.
11. b.
12. d.

Chapter 16: Drug licensing and pharmacopoeia

1. c.
2. a.
3. a.
4. a.
5. c.

Index